Cementing Technology and Procedures

Cementing Technology and Procedures

Editor

Shivraj Choudhary

Cementing Technology and Procedures

Edited by **Shivraj Choudhary**

Printed in 2017

ISBN: 978-1-68117-340-5

Library of Congress Control Number: 2015939252

© 2016 by
SCITUS Academics LLC,
616, Corporate Way, Suite 2, 4766,
Valley Cottage, NY 10989

www.scitusacademics.com

Contents

vi

Preface

Primary cementing is the process of placing cement in the annulus between the casing and the formations exposed to the wellbore. The major objective of primary cementing has always been to provide zonal isolation in the wellbore of oil, gas, and water wells to exclude fluids such as water or gas in one zone from oil in another zone. To achieve this objective, a hydraulic seal must be obtained between the casing and the cement. And between the cement and the formations, while at the same time preventing fluid channels in the cement sheath. This requirement makes primary cementing the most important operation performed on a well. Without complete zonal isolation in the wellbore. The well may never reach its full producing potential. Remedial work required to repair a faulty cementing job may do irreparable harm to the producing formation.

Editor

Formation of Platinum (Pt) Nanocluster Coatings on K-OMS-2 Manganese Oxide Membranes by Reactive Spray Deposition Technique (RSDT) for Extended Stability during CO Oxidation

Hector F. Garces[1], Justin Roller[2, 3], Cecil K. King'ondu[4], Saminda Dharmarathna[5], Roger A. Ristau[6], Rishabh Jain[2, 3], Radenka Maric[2, 3, 7], and Steven L. Suib[5, 6]

[1]School of Engineering, Brown University, Providence, USA

[2]Department of Materials Science & Engineering, University of Connecticut, Storrs, USA

[3]Center for Clean Energy Engineering, University of Connecticut, Storrs, USA

[4]Department of Sustainable Energy Science and Engineering, Nelson Mandela African Institution of Science and Technology, Arusha, Tanzania

[5]Department of Chemistry, University of Connecticut, Storrs, USA

[6]Institute of Materials Science, University of Connecticut, Storrs, USA

[7]Chemical and Biomolecular Engineering Department, University of Connecticut, Storrs, US

ABSTRACT

Nanocluster formation of a metallic platinum (Pt) coating, on manganese oxide inorganic membranes impregnated with multiwall carbon nanotubes (K-OMS-2/MWCNTs), applied by reactive spray deposition technology (RSDT) is discussed. RSDT applies thin films of Pt nanoclusters on the substrate; the thickness of the film can be easily controlled. The K-OMS-2/MWCNTs fibers were enclosed by the thin film of Pt. X-ray diffraction (XRD), scanning electron microscopy/X-ray energy dispersive spectroscopy (SEM/XEDS), focus ion beam/scanning electron microscopy (FIB/SEM), transmission electron microscopy (TEM), and X-ray 3D micro-tomography (MicroXCT) which have been used to characterize the resultant Pt/K-OMS-2/MWCNTs membrane. The non-destructive characterization technique (MicroXCT) resolves the Pt layer on the upper layer of the composite membrane and also shows that the membrane is composed of sheets superimposed into stacks. The nanostructured coating on the composite membrane material has been evaluated for carbon monoxide (CO) oxidation. The functionalized Pt/K-OMS-2/MWCNTs membranes show excellent conversion (100%) of CO to CO_2 at a lower temperature 200°C compared to the uncoated K-OMS-2/MWCNTs. Moreover, the Pt/K-OMS-2/MWCNTs membranes show outstanding stability, of more than 4 days, for CO oxidation at 200°C.

INTRODUCTION

Manganese oxide K-OMS-2 is a porous mixed-valent metal oxide with applications in catalysis [1-3], environmental remediation [4], sorption processes [5], and microbial fuel cells [6]. The rationale for the synthesis of this octahedral molecular sieve (OMS) is based on its low cost, processability, stability, and excellent catalytic activity in different redox reactions [7]. Its processability permits the preparation of K-OMS-2 manganese oxide materials composed of endless-type nanofibers that can readily be assembled into a paper or membrane [8]. This versatile free-standing structure in the form of a membrane is flexible, re-dispersible, foldable, moldable, and can be modified by ion-exchange, doping, distributed over large areas for clean-up as well as being used as a supporting structure to produce composite materials. Membranes are of special interest due to their porosity, permeability, and conductivity and hence their potential uses as sensors, catalysts, and in separation processes that remove bacteria, microorganisms, particulates, and organic material. Membranes can be assembled in several geometries facilitating the integration into different devices offering operational features unavailable with bulk materials [9]. Nano-composites of MWCNTs filled with MnO_2 have already been prepared and the results show improvements in both electrochemical and conduction properties [10]. The MWCNTs improve the connectivity between electrochemical isolated metallic centers and the manganese oxide nano-wires in the membrane, resulting in enhancement in the electron transport properties throughout the porous materials. Cryptomelane manganese oxide K-OMS-2 is a promising material to use in batteries and shows good capacity to store electrical charge while also being inexpensive and environmentally friendly. The addition of MWCNTs to K-OMS- 2 membranes is expected to improve the electrical, mechanical, and chemical resistance performance of the composite as well as to facilitate charge transfer.

Deposition of metals on K-OMS-2 membranes, in the open atmosphere, using a flame synthesis process has not been explored. This process constitutes a new cost-effective route to functionalize the inexpensive inorganic membrane and thereby incorporate new functionalities leading to improved performance in: capacitance, adsorption, and control over conversion and selectivity in catalysis.

Preparation of novel inorganic-organic composite materials and non-destructive characterization (NDC) has become very important in the field of materials science. NDC techniques permit the visualization of the component in load and the interaction with the host structure potentially reveals features commonly accessed only by time-demanding sample preparation procedures (i.e., isolation in epoxy followed by polishing). Additionally, lightweight materials with catalytic activity used as supports have attracted much attention due to their synergistic effects; they drive reactions towards higher yields and increase the interaction between the substrate and support.

In this work, a simple, fast, and new deposition technique has been employed for the formation of a conformal, homogeneous coating of Pt on inorganic K-OMS-2 impregnated with MWCNTs (i.e., the membrane substrate). The coated membrane (Pt/K-OMS-2/MWCNTs) showed excellent catalytic activity at a relatively low temperature (200°C). Thin conformal films of Pt were deposited using Reactive Spray Deposition Technology (RSDT). RSDT is a direct, dry flame based synthesis process and is configured to deposit directly onto a KOMS-2/MWCNT composite membrane [11-16]. RSDT is a method of depositing films through combustion of metal-organic or metal-inorganic compounds dissolved in a solvent, and has emerged as an analogue to other deposition techniques such as atomic layer deposition (ALD), chemical vapor deposition (CVD), pulse laser deposition (PLD), and physical vapor deposition (PVD). Some key advantages of the RSDT process over traditional processing methods are 1) a reduction in the number of processing steps required for catalyst formation, 2) avoiding wet processing routes, 3) lack of requirement for vacuum, 4) power consumption < 2 kW, 5) ease of stoichiometry control by mixing of precursors in the liquid phase, 6) no drying cycle, and 7) direct deposition. The process essentially combines the catalyst production and film formation steps into one, takes place in the open atmosphere and eliminates the need to dispose of solvent waste; the solvent is completely combusted to CO_2 and H_2O. Figure 1 (left) shows a schematic of the RSDT process. The figure on the right shows the deposition hardware set-up for thin film deposition.

Formation of nanocrystallites, via the RSDT technique, occurs through a sequential growth process that involves multiple intermediate steps. Nucleation occurs from the vapor phase whereby the metal or metal oxide grows into a primary particle during residence, along the length of the hot reactive zone [17]. There may be several pathways through which the vaporized metal reacts, nucleates, and grows either during time of flight or directly onto the substrate [18].

The precursor, once the droplet exits from the nozzle, proceeds through the following steps: heats up to the boiling point of the solvent; precipitates due to a rapid solvent shell volume decrease (i.e., simultaneous evaporation and combustion); decomposes; phase changes from solid to vapor; and finally undergoes a series of redox reactions. Pt^{2+} is reduced to Pt metal. Formation of the nanocrystallite particles, during time-of-flight, occurs prior to film formation through a multi-step process on a time scale of milliseconds. The general mechanism of particle growth, once the precursor has vaporized, occurs by: homogeneous reactions, nucleation, surface growth, cluster dynamics (a transitory state between single atoms and solid material), coalescence, aggregation, and agglomeration [19]. The solid particle passes through the following size classifications during the growth process: monomer formation, cluster, primary particle, nanoparticle, and then agglomerate [19-21]. Depending on the processing conditions, a film can form from the vapor phase (i.e., the product reaches the substrate at a stage somewhere between the monomer and nanoparticle pathway), either by a physical impingement of a fully formed nanoparticle (i.e., a ballistic collision), or by a combination of both mechanisms.

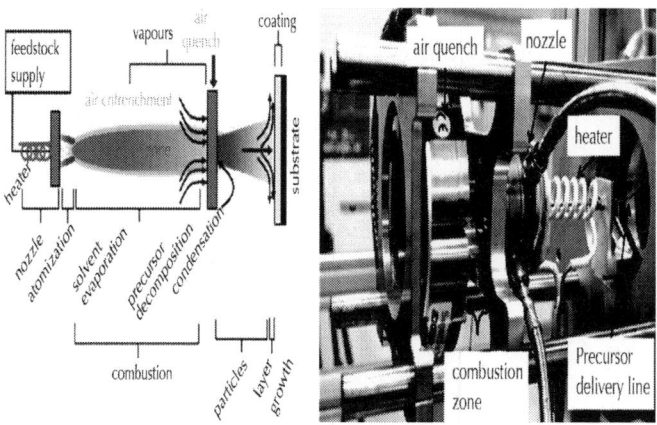

Figure 1: Schematic diagram of the RSDT process (left), and a mirror image of the process as set-up for deposition, substrate not shown (right).

The exact mechanism of growth is affected by the residence time in a given thermal profile, concentration of reactants, the precursor composition, oxidant/fuel flow rates, and the distance between the substrate and the nozzle. The thermal profile is controlled by the equivalence ratio, choice of fuel, quench distance, stand-off distance, flow rate, and nozzle design. Primary particle growth is arrested through rapid cooling using an air quench, to create a fast non-equilibrium phase change [22]. The time of flight, zone temperature profile, stand-off distance, and locations of the quench are critical to formation of the desired metal and morphology [19].

EXPERIMENTAL

Manganese Oxide Inorganic Membrane Synthesis

The inorganic K-OMS-2 membrane was prepared as previously reported [8]. Briefly, 11.37 g of potassium sulphate (K_2SO_4), 17.64 g of potassium persulphate ($K_2S_2O_8$), 7.35 g of manganese sulfate monohydrate ($MnSO_4 \cdot H_2O$), and 70 mL of distilled deionized water (DDW) are mixed together in a Teflon liner and placed in an autoclave at 250°C for 4 days in a conventional oven. After the synthesis, the material was dispersed in DDW. The K-OMS-2/MWCNTs composite membrane was formed after purification of the MWCNTs using an acid method procedure [23]. Briefly, the MWCNTs were boiled in concentrated (68%) HNO_3 at 120°C under reflux for about 48 h, washed with DDW, NH_4OH, DDW, HCl, DDW, filter, and finally dried. Finally, about 3.5 wt% of the purified MWCNTs was re-dispersed with the K-OMS-2 fibers to obtain the composite membrane.

Deposition of Pt on K-OMS-2 Membrane

Metallic platinum was deposited on modified K-OMS-2 membranes with MWCNTs by reactive spray deposition technology (RSDT). Briefly, the method involved pumping platinum acetylacetonate (PtAcac) dissolved in a toluene/propane solvent (20 wt% propane) through an atomizing nozzle at a flow rate of 4 mL/min and combusting the atomized spray (Figure 1). The deposition zone was directed onto a 4 mm × 4 mm K-OMS-2/ MWCNTs substrate occurred by impinging the flame on the substrate at a stand-off distance of 15 cm. The PtAcac was dissolved into toluene and further diluted with propane to a final metal concentration of 6.1 mm. The solution was heated in the atomizing nozzle via induction to a temperature of 170˚C. The solvent was passed through the nozzle at an average pressure of 190 psi, producing a flame approximately 7.5 cm long. A sheath of oxygen surrounded the atomized spray at a flow rate of 12 slpm. The average substrate temperature was 100˚C during the deposition.

K-OMS-2 Membrane Characterization

X-ray diffraction analysis was performed to evaluate the nature of the as prepared membrane, and after deposition of Pt on the modified membrane. The inorganic membranes were analyzed in an Ultima IV Rigaku X-ray diffractometer (Cu K radiation). Diffraction patterns were obtained in the range of 5 - 75 2 degrees at a scan rate of 2˚ min⁻¹. Scanning electron microscopy (SEM) was performed in an FEI Quanta ESEM 250 scanning electron microscope. The sample was cut in a square of about 5 mm and placed in a carbon tape. A cross section of the Pt deposited on the substrate was made after immersing a piece Pt/K-OMS-2/MWCNTs in an epoxide, and polishing with 1200, 600, and 320 grit grinding paper, and polishing with 5 um, 3 um, 1 um, and 0.3 um alumina powders, and 0.1 silica dispersion in a cloth sheet. Mapping of the samples were obtained by energy dispersive X-ray analysis (EDAX) in an FEI Quanta ESEM 250 operated at 20.0 kV with X-ray spectra acquired and processed with an Ametek Genesis Apex 4. High temperature scanning electron microscopy (HTSEM) was performed in an FEI Quanta ESEM 250 scanning electron microscope. Membrane type Pt/K-OMS-2/MWCNTs were heated to the analysis temperature at 50˚C min⁻¹ and stabilized for 30 minutes before an SEM image was taken. FIB/SEM images were acquired using The Strata 400 STEM DualBeam system equipped with Focused Ion Beam (FIB) technology and a Flipstage/STEM assembly. This system permits complete in situ sample preparation and high-resolution analysis. Transmission electron microscopy (TEM) micrographs were obtained using a JEOL JEM 2010 FasTEM operating at 200 kV. The specimens were loaded onto a carbon-coated gold grid. High resolution X-ray 3D tomography was performed in a MicroXCT- 400. The specifications for the tomography employed a 40 keV accelerating voltage and a beam current of 4 W. Two thousand projections were collected from −90 to 90 theta with an exposure time of 14 second per image. Reconstruction and visualization employed were performed afterwards.

CO Oxidation

The CO oxidation in a temperature range from 50°C to 400°C employed a vertical fixed-bed tubular reactor made of quartz with an internal diameter of 4 mm. In each experiment, the membrane was cut into square pieces of about 1 mm, packed in the quartz reactor, and held by quartz wool in both ends. The catalyst was heated to the analysis temperature with an Ar down flow of 40 CCM and held for 30 minutes before the beginning of the CO oxidation. Afterwards, three consecutive automatic injections were sampled at each evaluation temperature. The analytical system comprised an Agilent 3000 Micro GC equipped with two thermal conductivity detectors. A Molecular Sieve and a Plot Q capillary column were used for the separation. Tubing and fittings were stainless steel throughout. In each experiment, about 0.06 g of K-OMS-2/MWCNTs and Pt/K-OMS-2/ MWCNTs catalyst was placed in the reactor supported by quartz wool. A furnace with PID control held the reactor temperature constant. The thermocouple was placed at the top of the catalyst bed. Mass flow controllers (MFC) were used to control flow rates, feed, and composition. A certified gas mixture (Airgas; 10% CO in N_2), pure O_2 and Ar as balance were used for the CO oxidation. Stability tests for CO oxidation were performed at 200°C. For the stability test, the reactor containing the catalyst (Pt/K-OMS-2/MWCNTs) was ramped to 200°C, purged for 30 minutes with Argon down flow, and kept for about 4 days for the analysis. After the purging, 3 simultaneous sampling injections at different time intervals were directed into the Column and the outlet concentrations were averaged.

RESULTS AND DISCUSSION

X-Ray Diffraction

The diffraction patterns for the K-OMS-2/MWCNTs, and the Pt deposited membrane (Pt/K-OMS-2/MWCNTs) are presented in Figure 2. Diffraction patterns (a) for the impregnated K-OMS-2 with MWCNTs correspond to the tetragonal manganese oxide K-OMS-2 cryptomelane phase (ICSD No. 01-070-8072). Peaks for face centered cubic metallic Pt can be observed after deposition by RSDT [Figure 2(b) Pt/K-OMS-2/MWCNTs]. 1.17 wt % of metallic platinum was loaded in the composite membrane. Diffraction peaks for the platinum phase deposited on a bare aluminum stub as a control during the deposition by RSDT and X-ray diffraction patterns of cryptomelane manganese oxide and metallic platinum with a face centered cubic structure (ICDD No. 00-004-082) plotted using crystal maker software [24] are in accordance with the results (Supporting Information, Figures S1, S2).

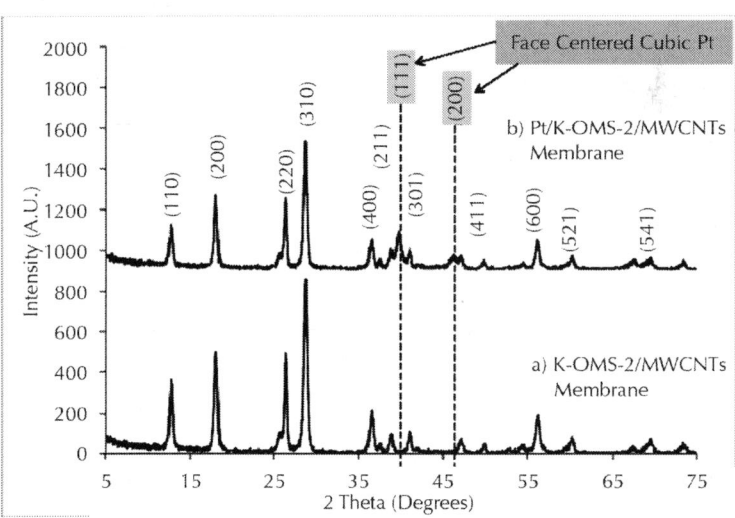

Figure 2: X-ray diffraction patterns of (a) fresh K-OMS-2/ MWCNTs and (b) Pt/K-OMS-2/MWCNTs. Diffractions peaks correspond to cryptomelane type manganese oxide (ICSD No. 01-070-8072) and metallic platinum face centered cubic (ICDD No. 00-004-082).

The Pt precursor deposited on the K-OMS-2/MWCNTs membrane material was completely and successfully transformed into crystalline metallic Pt coating as depicted by the X-ray diffraction pattern, Figure 2. The reflections that correspond to metallic platinum (111) and (200) are clearly observed at angles of about 39.5 and 46.0 degrees two-theta respectively, along with the manganese oxide K-OMS-2 cryptomelane type material XRD peaks, Figure 2(b). The platinum acetylacetonate (PtAcac) precursor was not detected after the deposition; this indicates a complete decomposition during the deposition. Additionally, the substrate temperature, due to the exothermic combustion from the torch, did not destroy the K-OMS-2 support. The substrate temperature was held constant at 100°C and the stand-off distance between the nozzle and substrate was held at 15 cm. The stability of the support after exposure to the reactive zone is confirmed by the post-deposition XRD pattern, which shows the pristine tetragonal phase of the K-OMS-2, (ICSD No. 01-070-8072). The interaction of the heat, from the flame, with the substrate does not induce structural changes; the crystal structure of the manganese oxide was preserved. Cryptomelane-type manganese oxide (K-OMS-2) materials are stable up to 800°C depending on the synthetic conditions under which they are prepared [25]. Therefore, unlike using the RSDT technique reported herein, loading active components on K-OMS-2 materials using high temperature deposition methods would certainly transform the cryptomelane type structure to more stable phases such as Mn_2O_3. Consequently, several physical properties, including its mixed-valence nature, would be altered and adversely affect catalytic activity. The substrate can be kept at relatively low temperatures (e.g. <150°C) by controlling the proximity of the torch to the substrate; this stand-off avoids unwanted phase transformations, or oxidation of components in the composite such as the carbon nanotubes. At the same time, the system is flexible allowing the possibility to deposit materials on a substrate maintained at high temperatures by an external heating device or nearing the torch during the deposition. At one extreme the luminous tip of the flame could be impinged directly onto the substrate creating temperatures around 800°C.

Scanning Electron Microscopy

Scanning electron microscopy (SEM) images are shown in Figure 3. Micrographs (a), (b) and (c), (d) are for the fresh K-OMS-2/MWCNTs membrane and Pt/K-OMS- 2/MW-CNTs after RSDT deposition of Pt, respectively. The membrane material is composed of individual fibers with different diameters, some of them as low as 20 nm and several microns in length that intertwined in different directions forming a porous blanket, Figures 3(a), (b). Surface fibers are homogeneously coated in Pt, by RSDT, showing an island growth morphology around the former K-OMS-2/MWCNTs composite membrane. Empty sections on the surface of the blanket of fibers are absent, Figures 3(c), (d). X-ray energy dispersive spectroscopy (XEDS) mapping of a cross section of the coated membrane and focus ion beam scanning electron microscopy (FIB/SEM) images are presented inFigure 4. Layers of superimposed K-OMS-2 that form a stack are evident in the cross sectional image, Figure 4(a). EDS mapping of the cross section shows a thick layer of Pt in the upper section of the membrane along with elements that constitute K-OMS-2 in the inner layers. The coating of metallic Pt that wraps the composite K-OMS-2/MWCNTs fibers is up to 60 nm thick, Figures 4(b), (c).

Figure 3: Scanning electron microscopy images of (a), (b) KOMS-2/MWCNTs before coating, and (c), (d) Pt/K-OMS- 2/MWCNTs after Pt deposition by RSDT.

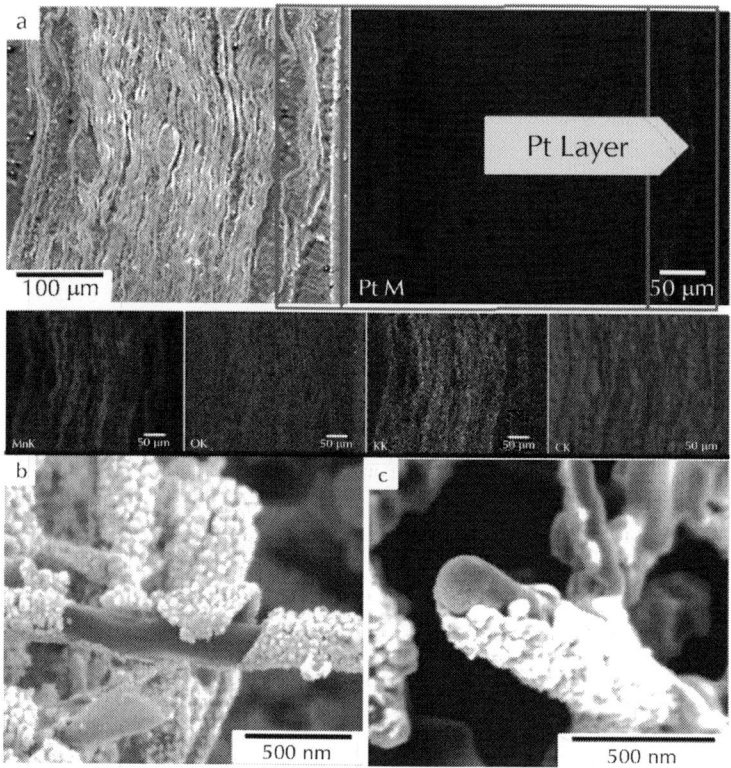

Figure 4: Scanning electron microscopy of (a) cross section of Pt/K-OMS-2/ MWCNTs and EDX mapping, and (b), (c) focus ion beam scanning electron microscopy (FIB/SEM) of coated Pt/KOMS-2/MWCNTs.

A uniform coating of Pt covered the upper fibers of K-OMS-2/ MWCNTs membrane, Figures 3(c),(d). The resultant conformal film of Pt nanoclusters covered the individual fibers on the upper-side of the membrane completely, without inducing any fiber aggregation. Moreover, the Pt coating did not cover the interstitial spaces between fibers substantially and this preserved the porous nature of the substrate (Supporting Information, Figure S3); this is a property critical to the catalytic performance. High temperature scanning electron microscopy (HTSEM) of a selected area also shows no change in morphology or sintering, agglomeration, and delamination of the coated Pt film with temperature. Possible decomposition of unreacted reagents after the deposition was also not observed (Supporting Information, Figure S4). This observation confirms that RSDT, which operates in the open atmosphere, can effectively be employed to functionally deposit controlled films on metal oxide inorganic membrane supports or composites without leaving undesirable by-products that might poison the functionalized oxide thereby decreasing their activity in their intended applications. Focused ion beam scanning electron microscopy (FIBSEM) revealed that the Pt film is constructed of small clusters of less than 50 nm in size that form a shell over the manganese oxide fibers and carbon nanotubes as shown by the FIBSEM micrographs in Figures 4(b), (c), that show a debonded Pt film after FIB sectioning. The Pt coat would, otherwise, not come apart after being deposited and would not leave the fibers exposed in the upper layer of the K-OMS-2/MWCNTs membrane. Additionally, deposition by RSDT produces a conformal film, which indicates the capability of the deposition method to permeate porous structures thereby facilitating the complete coverage of the substrate.

X-Ray 3D Tomography

X-ray 3D tomography resolved the layer of platinum deposited on the K-OMS-2/MWCNTs membrane, Figure 5. It is also observed, as by SEM imaging of a cross section, that the membrane is composed of superimposed layered sheets that form a stack. The X-ray absorption levels of the different components of the composite system define the structure. The contrast between the high (Pt) and the low absorption manganese material in the membrane are well resolved. There is no dispersion of Pt in the inorganic matrix, only the upper manganese oxide layer got covered by the nanostructured array of metallic Pt, Figure 5(a). The orthogonal view of the 3D image also illustrates the layered structure of the inorganic matrix, Figures 5(b)-(d).

Resolving the layered nature of the Pt/K-OMS-2/ MWCNTs membrane involves destructive techniques with high energy Argon ions beam (FIB/SEM) and/or other time consuming procedures that might involve embedding the material in epoxies that go through a time consuming polishing procedure to access the material in order to image it (Figure 4(a)). Unlike in the above-mentioned techniques, this work presents a non-destructive technique that not only resolves the upper layer of the coated material but also permit the visualization of the stacking layers of the inorganic constituent, K-OMS-2, that forms the composite membrane material. Considering the attenuation of X-rays by dense bodies, carbon (C, Z = 6) cannot be accessed. However, the contrast between the high atomic number Pt material (Pt, Z = 78) deposited, and the low atomic number elements that constitute the layered manganese oxide membrane (Mn, Z = 25; K, Z = 19; O, Z = 8) are well resolved.

The use of X-ray tomography in materials science has been recently applied to study the 3D microstructural interactions in materials [26-31], and has become an important characterization technique with three-dimensional output information about the embedded phase, size, shape, and orientation.

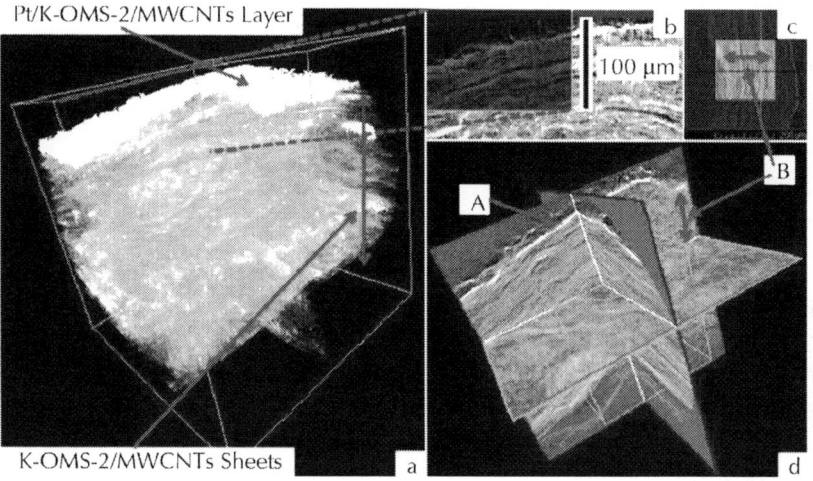

Figure 5: X-ray 3D micro tomography of (a) Pt/K-OMS-2/ MWCNTs and layer of inorganic K-OMS-2 in the membrane and (b), (c) Pt/K-OMS-2/MWCNTs layers, (d) three orthogonal slices of Pt/K-OMS-2/MWCNTs showing Pt/ KOMS-2/MWCNTs (A), and K-OMS-2 layers (B).

Figure 5(a) presents the X-ray 3D tomography for the Pt/K-OMS-2/ MWCNTs showing the layer of Pt deposited on the membrane material and the stacking of layers of sheets that make up the membrane. The 3D orthogonal view (Figure 5(d)) shows wavy lines along each orthogonal plane merging to form the stack of the layers. This can also be observed in Figure 5(b). The resolution of the instrument used for the 3D tomography (~0.5 μm) does not resolve the nanometer size of the individual fibers that compose the composite membrane but presents a clear picture at the micron scale of the organization of the material inside the membrane. The X-ray absorption levels of the components attenuated on the composite define the structure. There is no dispersion of Pt in the inorganic K-OMS-2 matrix; only the upper layer was covered by the nanostructured array of metallic Pt, Figure 5(a). Although the membrane is permeated by Pt to form the conformal structure, this permeation only penetrates into a few upper layers. Increases in the thickness of the coated material, in the upper layer, with deposition time eventually reduce the permeation to the lower layers. This closes paths for the platinum and restricts access to layers of K-OMS-2/MWCNTs underneath; the functionalized Pt material is therefore restricted to a few microns at the surface of the substrate. This technique is therefore self-limiting in terms of substrate penetration. The purpose of the tomographic analysis is to highlight the internal microstructure of the substrate; the coated layer; and to gain an understanding of the interaction between the support and the host structure. The deposition is homogeneous along the support structure.

Transmission Electron Microscopy

Further morphological and crystal structure studies by transmission electron microscopy (TEM) are presented in Figure 6. Low magnification TEM image show K-OMS- 2 fibers and carbon nanotubes merged together and both hold platinum nanoparticles during the RSDT coating, Figures 6(a), (b). Individual fibers of K-OMS-2 with diameters of about 20 nm and length extending to several microns with some of them sticking together forming a bundle were observed, Figure 6(b). A high-resolution TEM (HRTEM) micrograph (Figure 6(c)) shows that the K-OMS-2 nanofibers are single crystalline with periodic lattice fringes of 3.0 Å corresponding to the inter-planar spacing of (310) planes. In Figure 6(e), HRTEM image shows a MWCNT hosting a coated Pt particle. The coated Pt particle shows lattice fringes of 2.3 Å corresponding to the (111) planar spacing for metallic platinum. The inter-planar spacings observed are all in accordance with those measured with XRD for the bulk powdered sample for both the K-OMS-2 and metallic Pt materials.

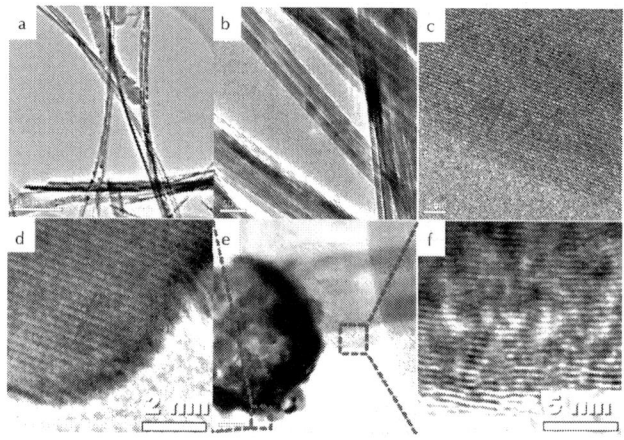

Figure 6: Transmission electron microscopy of (a) K-OMS- 2 fibers and carbon nanotubes, (b) fibers of K-OMS-2, (c) K-OMS-2 with periodic lattice fringes of 3.0 Å corresponding to the inter-planar spacing of (310) planes, (d) Pt particle hosted by a MWCNTs with lattice fringes of 2.3 Å corresponding to the (111) planar spacing for metallic platinum, (e) MWCNTs hosting the Pt particles, and (f) lattice fringes of MWCNTs.

As shown by HRTEM, Pt nanoparticles are distributed on both the K-OMS-2 fibers and MWCNTs,Figure 6(a). The coverage of the Pt film in the TEM micrographs is not homogeneous since the sample preparation procedure to obtain HRTEM images of the Pt/K-OMS-2/ MWCNTs membrane involved cutting the sample with a diamond blade, sonication, and depositing the sample onto a copper grid. As a result, many of the Pt particles peeled off from the surface of the K-OMS-2/MWCNTs fibers during this procedure. However, some of the platinum particles that do remain exhibit a size of about 15 nm (Figure 6(a); Supporting Information, Figure S5).

Carbon Monoxide (CO) Oxidation

Carbon monoxide (CO) oxidation is presented in Figure 7 in the range 50°C - 400°C. Lower temperatures (≤200°C) were found be optimal for high CO conversion with the coated Pt/K-OMS-2/MWCNTs material than the uncoated K-OMS-2/MWCNTs, reaching 100% conversion at 200°C. Higher temperatures (≥300°C) resulted in 100% conversion for both materials. A stability test at 200°C for CO oxidation shows that the coated Pt/K-OMS-2/MWCNTs remains active without apparent activity decrease for more than 4 days (Figure 7(b)).

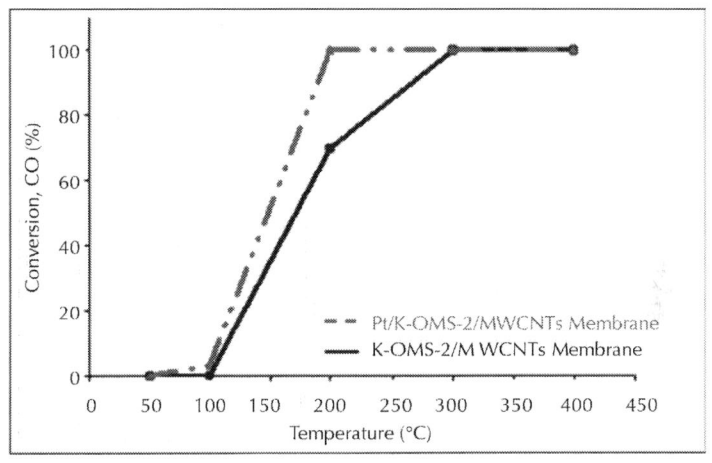

(a)

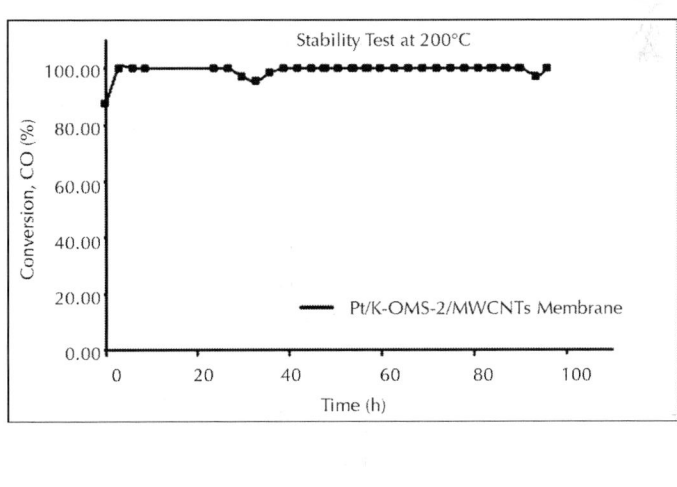

(b)

Figure 7. Carbon monoxide (CO) oxidation for (a) coated platinum Pt/K-OMS-2/MWCNTs and uncoated K-OMS-2/ MWCNTs, (b) stability test at 200°C for Pt/K-OMS-2/MWCNTs during 96 h.

Carbon monoxide oxidation is catalyzed by a wide variety of materials [32-41]. Several of these materials are supported on high temperature ceramics to impart mechanical, thermal resistance, and to avoid depletion of the active metal. Platinum is one of the most commonly used metals for CO oxidation. Pt is stable, resistant to moisture, and highly active for the formation of CO_2 [32,33,35]. The use of membranes is limited for these purposes in part because they cannot withstand larger pressure levels, limiting operational flexibility, and the use of higher pressures and in some cases the upper temperature. However, low drop pressures, flexibility, and permeation are some of the advantages that these structures can offer. The loading of Pt on K-OMS-2/MWCNTs composite membranes reflects the versatility in using a composite structure; these options incorporate different materials that impart flexibility in the membrane design and enhance other properties such as conduction. Pt coated membrane, Pt/K-OMS-2/MWCNTs showed higher activity than the uncoated K-OMS-2/MWCNTs, reaching 100% conversion not only at low temperature (200°C) but also being highly stable over a long period of time at 200°C, with no apparent degradation in catalytic conversion.

CONCLUSIONS

A simple, fast, and new deposition technique has been employed for the formation of highly homogeneous conformal Pt coatings on an inorganic K-OMS-2/MWCNTs composite membrane. The deposition of Pt obtained by RSDT, reduces the number of processing steps required for catalyst formation into one step in the open atmosphere, and eliminates the solvent disposal problem. Additionally, RSDT is a controllable carrier concentration, where the thickness of the layer can be built up from the precursor solution. The resultant Pt/K-OMS-2/MWCNTs membrane has been used for the CO oxidation over a wide temperature range (RT-400°C) without degradation of the coating. The functionalized membrane showed great performance and stability (>4 days) at lower temperatures than the uncoated one. This work also presents an alternative use for K-OMS-2 in the form of membranes incorporating this composite in catalytic processes. The work also presented a non-destructive characterization technique for the visualization of the high X-ray absorbing metallic phase supported in the nanomaterial K-OMS-2 phase. The technique resolved the coating layer and the nature of the inorganic K-OMS-2 membrane with minimal time-consuming sample preparation procedures.

ACKNOWLEDGEMENTS

We thank the Chemical Sciences, Geosciences, and Biosciences Division, Office of Basic Energy Sciences, Office of Sciences of the US Department of Energy for support of this work.

REFERENCES

1. C. K. King'ondu, N. Opembe, C. Chen, K. Ngala, H. Huang, A. Iyer, H. F. Garces and S. L. Suib, "Manganese Oxide Octahedral Molecular Sieves (OMS-2) Multiple Framework Substitutions: A New Route to OMS-2 Particle Size and Morphology Control," Advanced Functional Materials, Vol. 21, No. 2, 2011, pp. 312-323.http://dx.doi.org/10.1002/adfm.201001020

2. A. Iyer, J. Del-Pilar, C. K. King'ondu, E. Kissel, H. F. Garces, H. Huang, A. M. El-Sawy, P. K. Dutta and S. L. Suib, "Water Oxidation Catalysis Using Amorphous Manganese Oxides, Octahedral Molecular Sieves (OMS-2), and Octahedral Layered (OL-1) Manganese Oxide Structures," Journal of Physical Chemistry C, Vol. 116, No. 10, 2012, pp. 6474-6483.http://dx.doi.org/10.1021/jp2120737

3. S. Dharmarathna, C. K. King'ondu, W. Pedrick, L. Pahalagedara and S. L. Suib, "Direct Sonochemical Synthesis of Manganese Octahedral Molecular Sieve (OMS-2) Nanomaterials Using Cosolvent Systems, Their Characterization, and Catalytic Applications," Chemistry of Materials, Vol. 24, No. 4, 2012, pp. 705-712. http://dx.doi.org/10.1021/cm203366m

4. M. Abecassis-Wolfovich, R. Jothiramalingam, M. V. Landau, M. Herskowitz, B. Viswanathan and T. K. Varadarajan, "Cerium Incorporated Ordered Manganese Oxide OMS-2 Materials: Improved Catalysts for Wet Oxidation of Phenol Compounds," Applied Catalysis B: Environmental, Vol. 59, 2005, pp. 91-98. http://dx.doi.org/10.1016/j.apcatb.2005.01.001

5.   B. Hu, C. Chen, S. J. Frueh, L. Jin, R. Joesten and S. L. Suib, "Removal of Aqueous Phenol by Adsorption and Oxidation with Doped Hydrophobic Cryptomelane-Type Manganese Oxide (K-OMS-2) Nanofibers," Journal of Physical Chemistry C, Vol. 114, 2010, pp. 9835-9844. http://dx.doi.org/10.1021/jp100819a

6.   X. Li, B. Hu, S. L. Suib, Y. Lei and B. Li, "Electricity Generation in Continuous Flow Microbial Fuel Cells (MFCs) with Manganese Doxide (MnO_2) Cathodes," Biochemical Engineering Journal, Vol. 54, No. 1, 2011, pp. 10-15.http://dx.doi.org/10.1016/j.bej.2011.01.001

7.   S. L. Suib, "Porous Manganese Oxide Octahedral Molecular Sieves and Octahedral Layered Materials," Accounts of Chemical Research, Vol. 41, No. 4, 2008, pp. 479-487.http://dx.doi.org/10.1021/ar7001667

8.   J. Yuan, K. Laubernds, J. Villegas, S. Gomez and S. L. Suib, "Spontaneous Formation of Inorganic Paper-Like Materials," Advanced Materials, Vol. 16, No. 19, 2004, pp. 1729-1732. http://dx.doi.org/10.1002/adma.200400659

9.   J. A. Rogers, M. G. Lagally and R. G. Nuzzo, "Synthesis, Assembly and Applications of Semiconductor Nanomembranes," Nature, Vol. 477, No. 7362, 2011, pp. 45-53.http://dx.doi.org/10.1038/nature10381

10. S. I. A. Razak, S. H. S. Zein and A. L. Ahmad, "MnO_2- Filled Multiwalled Carbon Nanotube/Polyaniline Nanocomposites: Properties and Its Percolation Threshold," Nano: Brief Reports and Reviews, Vol. 6, No. 1, 2011, pp. 81-91.http://dx.doi.org/10.1142/S1793292011002378

11. J. Roller, R. Neagu, F. Orfino and R. Maric, "Supported and Unsupported Platinum Catalysys Prepared by a OneStep Dry Deposition Method and Their Oxygen Reduction Reactivity in Acidic Media," Journal of Materials Science, Vol. 47, No. 11, 2012, pp. 4604-4611. http://dx.doi.org/10.1007/s10853-012-6324-3

12. R. Maric, J. Roller and R. Neagu, "Flame-Based Technologies and Reactive Spray Deposition Technology for Low-Temperature Solid Oxide Fuel Cells: Technical and Economic Aspects," Journal of Thermal Spray Technology, Vol. 20, No. 4, 2011, pp. 696-718. http://dx.doi.org/10.1007/s11666-011-9645-x

13. R. Maric, K. Furusaki, D. Nishijima and R. Neagu, "Thin Film Low Temperature Solid Oxide Fuel Cell (LTSOFC) by Reactive Spray Deposition Technology (RSDT)," ECS Transactions, Vol. 35, No. 1, 2011, pp. 473-481.

14. R. Maric, R. Neagu, Y. Zhang-Steenwinkel, F. Van Berkel and B. Rietveld, "Reactive Spray Deposition Technology—An One-Step Deposition Technique for Solid Oxide Fuel Cell Barrier Layers," Journal of Power Sources, Vol. 195, No. 24, 2010, pp. 8198-8201.http://dx.doi.org/10.1016/j.jpowsour.2010.06.053

15. R. Nédélec, R. Neagu, S. Uhlenbruck, R. Maric, D. Sebold, H. Buchkremer and D. Stöver, "Gas Phase Deposition of Diffusion Barriers for Metal Substrates in Solid Oxide Fuel Celss," Surface & Coatings Technology, Vol. 205, No. 16, 2011, pp. 3999-4004. http://dx.doi.org/10.1016/j.surfcoat.2011.02.021

16. J. Roller, "Low Platinum Electrodes for Proton Exchange Fuels Cells Manufactures by Reactive Spray Deposition Technology," MASc Thesis, University of British Columbia, Vancouver, 2009.

17. A. Camenzind, W. R. Caseri and S. E. Pratsinis, "FlameMade Nanoparticles for Nanocomposites," Nano Today, Vol. 5, No. 1, 2010, pp. 48-65.http://dx.doi.org/10.1016/j.nantod.2009.12.007

18. T. T. Kodas and M. J. Hampden-Smith, "Aerosol Processing of Materials," Wiley-VCH, New York, 1999.

19. P. Roth, "Particle Synthesis in Flames," Proceedings of the Combustion Institute, Vol. 31, No. 2, 2007, pp. 1773- 1788. http://dx.doi.org/10.1016/j.proci.2006.08.118

20. R. Strobel and S. E. Pratsinis, "Flame Aerosol Synthesis of Smart Nanostructured Materials," Journal of Materials Chemistry, Vol. 17, No. 45, 2007, pp. 4743-4756.http://dx.doi.org/10.1039/b711652g

21. M. S. Wooldridge, "Gas-Phase Combustion Synthesis of Particles," Progress in Energy and Combustion Science, Vol. 24, No. 1, 1998, pp. 63-87.http://dx.doi.org/10.1016/S0360-1285(97)00024-5

22. K. Wegner and S. E. Pratsinis, "Nozzle-Quenching Process for Controlled Flame Synthesis of Titania Nanoparticles," AIChE Journal, Vol. 49, No. 7, 2003, pp. 1667-1675.http://dx.doi.org/10.1002/aic.690490707

23. L. Stobinski, B. Lesiak, L. Kover, J. Toth, S. Biniak, G. Trykowski and J. Judek, "Multiwall Carbon Nanotubes Purification and Oxidation by Nitric Acid Studied by the FTIR and Electron Spectroscopy Methods," Journal of Alloys and Compounds, Vol. 501, No. 1, 2010, pp. 77-84. http://dx.doi.org/10.1016/j.jallcom.2010.04.032

24. CrystalMaker Software Ltd., "CrystalMaker: A Crystal and Molecular Structure Program for Mac and Windows," Oxford, Version 2.1.5, 1994-2009. http://www.crystalmaker.com/

25. R. N. DeGuzman, Y. Shen, E. J. Neth, S. L. Suib, C. O'Young, S. Levine and J. M. Newsam, "Synthesis and Characterization of Octahedral Molecular Sieves (OMS-2) Having the Hollandite Structure," Chemistry of Materials, Vol. 6, No. 6, 1994, pp. 815-821.http://dx.doi.org/10.1021/cm00042a019

26. E. Maire, N. Gimenez, V. Sauvant-Maynot and H. Sauterean, "X-Ray Tomography and Three-Dimensional Image Analysis of Epoxy-Glass Syntactic Foams," Philosophical Transactions of the Royal Society A, Vol. 364, No. 1838, 2006, pp. 69-88.

27. Q. Zhang, P. D. Lee, R. Singh, G. Wu and T. C. Lindley, "Micro-CT Characterization of Structural Features and Deformation Behavior of Fly Ash/Aluminum Syntactic Foam," Acta Materialia, Vol. 57, No. 10, 2009, pp. 3003- 3011.http://dx.doi.org/10.1016/j.actamat.2009.02.048

28. J. Kastner, B. Harrer, G. Requena and O. Brunke, "A Comparative Study of High Resolution Cone Beam XRay Tomography and Synchrotron Tomography Applied to Feand Al-Alloys," NDT&E International, Vol. 43, No. 7, 2010, pp. 599-605.http://dx.doi.org/10.1016/j.ndteint.2010.06.004

29. L. Salvo, M. Suery, A. Marmottant, N. Limodin and D. Bernard, "3D Imaging in Material Science: Application of X-Ray Tomography," Comptes Rendus Physique, Vol. 11, No. 9, 2010, pp. 641-649. http://dx.doi.org/10.1016/j.crhy.2010.12.003

30. R. Moreno-Atanasio, R. A. Williams and X. Jia, "Combining X-Ray Microtomography with Computer Simulation for Analysis of Granular and Porous Materials," Particuology, Vol. 8, No. 6, 2010, pp. 81-99. http://dx.doi.org/10.1016/j.partic.2010.01.001

31. J. Kastner, B. Harrer and H. P. Degischer, "High Resolution Cone Beam X-Ray Computed Tomography of 3DMicrostructures of

Cast Al-Alloys," Materials Characterization, Vol. 62, No. 1, 2011, pp. 99-107. http://dx.doi.org/10.1016/j.matchar.2010.11.004

32.   C. Kwak, T. Park and D. J. Suh, "Preferential Oxidation of Carbon Monoxide in Hydrogen-Rich Gas over Platinum-Cobalt-Alumina Aerogel Catalysts," Chemical Engineering Science, Vol. 60, No. 5, 2005, pp. 1211-1217. http://dx.doi.org/10.1016/j.ces.2004.07.126

33.   D. J. Suh, C. Kwak, J. Kim, S. M. Kwon and T. Park, "Removal of Carbon Monoxide from Hydrogen Rich Fuels by Selective Low-Temperature Oxidation over Base Metal Added Platinum Catalysts," Journal of Power Sources, Vol. 142, No. 1, 2005, pp. 70-74.http://dx.doi.org/10.1016/j.jpowsour.2004.09.012

34.   H. Igarashi, H. Uchida, M. Suzuki, Y. Sasaki and M. Watanabe, "Removal of Carbon Monoxide from Hydrogen-Rich Fuels by Selective Oxidation over Platinum Catalysts Supported on Zeolite," Applied Catalysis A: General, Vol. 159, 1997, pp. 159-169.http://dx.doi.org/10.1016/S0926-860X(97)00075-6

35.   K. Teruuchi, H. Habazaki, A. Kawashima, K. Asami and K. Hashimoto, "Amorphous Nickel-Base Alloy Catalysts for Oxidation of Carbon Monoxide by Oxygen and Nitrogen Monoxide," Applied Catalysis, Vol. 76, No. 1, 1991, pp. 79-93. http://dx.doi.org/10.1016/0166-9834(91)80006-I

36.   E. M. C. Alayon, J. Singh, M. Nachtegaal, M. Harfouche and J. A. Van Bokhoven, "On Highly Active Partially Oxidized Platinum in Carbon Monoxide Oxidation over Supported Platinum Catalysts," Journal of Catalysis, Vol. 263, No. 2, 2009, pp. 228-238.http://dx.doi.org/10.1016/j.jcat.2009.02.010

37.   G. Gürdag and T. Hahn, "The Oxidation of Carbon Monoxide on Platinum-Supported Binary Oxide Catalysts," Applied Catalysis A, Vol. 192, No. 1, 2000, pp. 51-55.http://dx.doi.org/10.1016/S0926-860X(99)00332-4

38.   D. Gavril, N. A. Katsanos and G. Karaiskakis, "Gas Chromatographic Kinetic Study of Carbon Monoxide Oxidation over Platinum-Rhodium Alloy Catalysts," Journal of Chromatography A, Vol. 852, No. 2, 1999, pp. 507-523.http://dx.doi.org/10.1016/S0021-9673(99)00642-1

39.   M. Stancheva, S. Manev, D. Lazarov and M. Mitov, "Catalytic Activity of Nickel Based Amorphous Alloys for Oxidation

of Hydrogen and Carbon Monoxide," Applied Catalysis A: General, Vol. 135, No. 1, 1996, pp. L19-L24. http://dx.doi.org/10.1016/0926-860X(95)00275-8

40.   K. Wu, Y. Tung, Y. Chen and Y. Chen, "Catalytic Oxidation of Carbon Monoxide over Gold/Iron Hydroxide Catalysts at Ambient Conditions," Applied Catalysis B: Environmental, Vol. 53, No. 2, 2004, pp. 111-116. http://dx.doi.org/10.1016/j.apcatb.2004.05.008

41.   P. V. Gosavi and R. B. Biniwale, "Catalytic Preferential Oxidation of Carbon Monoxide over Platinum Supported on Lanthanum Ferrite-Ceria Catalysts for Cleaning Hydrogen," Journal of Power Sources, Vol. 222, 2013, pp. 1-9.http://dx.doi.org/10.1016/j.jpowsour.2012.07.095

Alkaline and Alkaline-Earth Ceramic Oxides for CO$_2$ Capture, Separation and Subsequent Catalytic Chemical Conversion

Margarita J. Ramírez-Moreno[1, 2], Issis C. Romero-Ibarra[1], José Ortiz-Landeros[2], and Heriberto Pfeiffer[1]

[1]Instituto de Investigacios en Materiales, Universidad Nacional Autónoma de México, Circuito exterior s/n, Ciudad Universitaria, Del. Coyoacán, México DF, Mexico

[2]Departamento de Ingeniería en Metalurgia y Materiales, Escuela Superior de Ingeniería Química e Industrias Extractivas, IPN, UPALM, México DF, Mexico

INTRODUCTION

The amounts of anthropogenic carbon dioxide (CO$_2$) in the atmosphere have been raised dramatically, mainly due to the combustion of different carbonaceous materials used in energy production, transport and other important industries such as cement production, iron and steelmaking. To solve or at least mitigate this environmental problem, several alternatives have been proposed. A promising alternative for reducing the CO_2 emissions is the separation and/or capture and concentration of the gas and its subsequent chemical transformation. In that sense, a variety of materials have been tested containing alkaline and/or alkaline-earth oxide ceramics and have been found to be good options.

The aforementioned ceramics are able to selectively trap CO_2 under different conditions of temperature, pressure, humidity and gas mixture composition. The influence of those factors on the CO_2 capture (physically or chemically) seems to promote different sorption mechanisms, which depend on the material's chemical composition and the sorption conditions used. Actually, this capture performance suggests the feasibility of these kinds of solid for being used with different capture technologies and processes, such as: pressure swing adsorption (PSA), vacuum swing adsorption (VSP), temperature swing adsorption (TSA) and water gas shift reaction (WGSR). Therefore, the fundamental study regarding this matter can help to elucidate the whole phenomena in order to enhance the sorbents' properties.

CO$_2$ CAPTURE BY DIFFERENT ALKALINE AND ALKALINE-EARTH CERAMICS

Among the alkaline and/or alkaline-earth oxides, various lithium, sodium, potassium, calcium and magnesium ceramics have been proposed for CO_2 capture through adsorption and chemisorption processes [1-20]. These materials can be classified into two large groups: dense and porous ceramics. Dense ceramics mainly trap CO_2 chemically: the CO_2 is chemisorbed. Among these ceramics, CaO is the most studied one. It presents very interesting sorption capacities at high temperatures (T ≥ 600 °C). In addition to this material, alkaline ceramic oxides have been considered as possible captors, mostly lithium and sodium based ceramics (Li_5AlO_4 and Na_2ZrO_3, for example). In these cases, one of the most interesting properties is related to the wide temperature range in which some of these ceramics trap CO_2 (between 150 and 800 °C), as well as their high CO_2 capture capacity.

In these ceramics, the CO_2 capture occurs chemically, through a chemisorption process. At a micrometric scale, a general reaction mechanism has been proposed, where the following steps have been established: Initially, CO_2 reacts at the surface of the particles, producing the respective alkaline or alkaline-earth carbonate and in some cases different secondary phases. Some examples are:

$$Li_5AlO_4 + 2\,CO_2 \longrightarrow 2Li_2CO_3 + LiAlO_2$$

(1)

$$Na_2ZrO_3 + CO_2 \longrightarrow Na_2CO_3 + ZrO_2$$

(2)

$$CaO + CO_2 \longrightarrow CaCO_3$$

(3)

The above reactions show that surface products can be composed of carbonates, but as well they can contain metal oxides or other alkaline/alkaline-earth ceramics. The presence of these secondary phases can modify (improve or reduce) the diffusion processes described below [1].

Once the external carbonate shell is formed, different diffusion mechanisms have to be activated in order to continue the CO_2 chemisorption, through the particle bulk. Some of the diffusion processes correspond to the CO_2 diffusion through the mesoporous external carbonate shell, and some others such as the intercrystalline and grain boundary diffusion processes [1, 18, 21].

Figure 1 shows the theoretical CO_2 chemisorption capacities (mmol of CO_2 per gram of ceramic) for the most studied alkaline and alkaline-earth ceramics. As it can be seen, metal oxides (Li_2O, MgO and CaO) are among the materials with the best CO_2 capture capacities.

Nevertheless, Li_2O and MgO have not been really considered as possible options due to reactivity and kinetics factors, respectively. On the contrary, CaO is one of the most promising alkaline-earth based materials, with possible real industrial applications. Other interesting materials are ceramics with lithium or sodium phases, which present better thermal stabilities and volume variations than CaO. In addition, the sodium phases may present another advantage if the CO_2 capture is produced in the presence of steam. Under these conditions the sodium phases may produce sodium bicarbonate ($NaHCO_3$) as the carbonated phase, which is twice the amount of CO_2 could be trapped in comparison to the Na_2CO_3 product.

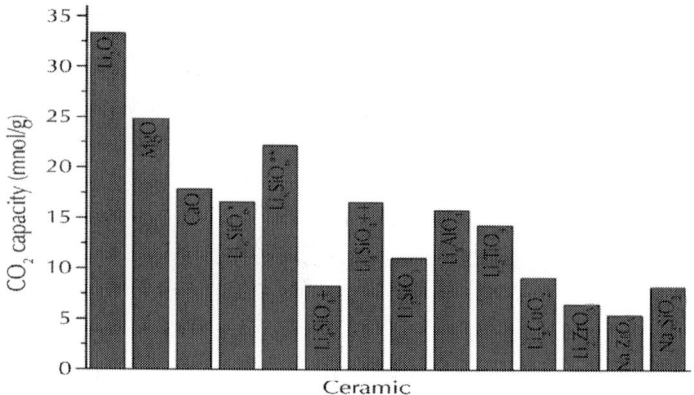

Figure 1: Theoretical CO_2 capture capacities for different alkaline and alka-line-earth ceramics. In the Li_8SiO_6(labeled as *) and Li_4SiO_4 (labeled as +), the maximum capacity can depend on the CO_2 moles captured in each different phase formed ($Li_8SiO_6 + CO_2 \rightarrow Li_4SiO_4 + CO_2 \rightarrow Li_2SiO_3 + Li_2CO_3$).

Other ceramics containing alkaline-earth metals are the layered double hydroxides (LDH) or hydrotalcite-like compounds (HTLc). LDHs, also called anionic clays due to their layered structure and structural resemblance to a kind of naturally-occurring clay mineral. These materials are a family of anionic clays that have received much attention in the past decades because of their numerous applications in many different fields, such as antacids, PVC additives, flame retardants and more recently for drug delivery systems and as solid sorbents of gaseous pollutants [22-24]. The LDH structure is based on positively charged brucite like [$Mg(OH)_2$] layers that consist of divalent cations surrounded octahedrally by hydroxide ions. These octahedral units form infinite layers by edge sharing [25]. Due to the fact that certain fraction of the divalent cations can be substituted by trivalent cations at the centers of octahedral sites, an excess of positive charge is promoted. The excess of positive charge in the main layers of LDHs is compensated by the intercalation of anions in the hydrated interlayer space, to form the three-dimensional structure. These materials have relatively weak bonds between the interlayer and the sheet, so they exhibit excellent ability to capture organic or inorganic anions. The materials are easy to synthesize by several methods such as co-precipitation, rehydration-reconstruction, ion exchange, hydrothermal, urea hydrolysis and sol gel, although not always as a pure phase [26].

The LDH materials are represented by the general formula:

$[M^{II}_{1-x}M^{III}_x(OH)_2]^{x+}[A^{m-}]_{x/m} \cdot nH_2O$

where M^{II} and M^{III} are divalent (Mg^{2+}, Ni^{2+}, Zn^{2+}, Cu^{2+}, etc.) and trivalent cations (Al^{3+}, Fe^{3+}, Cr^{3+}, etc.), respectively, and A^{m-} is a charge compensating anion such as CO_3^{2-}, SO_4^{2-}, NO_3^{-}, Cl^{-}, OH^{-}, where x is equal to the molar ratio of $[M^{III}/(M^{II} + M^{III})]$. Its value is commonly between 0.2 and 0.33, i.e., the M^{II}/M^{III} molar ratio is in the range of 4 - 2 [25], but this is not a limitation ratio and it depends on the M^{II} and M^{III} composition [27-29].

Among various CO_2 mesoporous adsorbents, LDH-base materials have been identified as suitable materials for CO_2 sorption at moderate temperatures (T ≤ 400 °C) [30-46] due to their properties such as large surface area, high anion exchange capacity (2-3 meq/g) and good thermal stability [37-40]. The LDH materials themselves do not possess any basic sites. For that reason, it is preferred to use their derived mixed oxides, formed by the thermal decomposition of LDH, which do exhibit interesting basic properties. Thermal decomposition of the material occurs in three stages, first at temperatures lower than 200 °C, at which the dehydration of superficial and interlayer water molecules takes place on the material. Then the second decomposition stage takes place in the range of 300-400 °C, at which the structure collapses due to a partial dehydroxylation process, typically associated with both the decomposition of Al-OH and the Mg-OH hydroxides. During dehydroxylation, changes occur in the structure. A portion of the trivalent cations of the brucite like layers migrates to the interlaminar region, allowing the preservation of the laminar characteristics of the material [41]. Finally, the total decomposition of the material occurs at temperatures higher than 400 °C, when the decarbonation process is completed [42].

Once the temperature reaches about 400 °C, LDH forms a three-dimensional network of compact oxygen with a disordered distribution of cations in the interstices, where the cations M^{+3} are tetrahedrally coordinated (interlayer region) and M^{+2} are octahedrally coordinated. The compressive-expansion stresses associated with the formation of the amorphous three-dimensional networks and their connection to the octahedral layer increases the surface area and pore volume, which can help improve the storage capacity properties, for example for gas sorption related applications, besides decreasing the ability of the Mg^{+2} cation to favor physisorption instead of chemisorption [30, 42]. For instance, the thermal evolution of the $Mg/Al\text{-}CO_3$ LDH structure is considered to be crucial in determining the CO_2 adsorption capacity, so there are several studies about this issue [42-44].

Reddy et al. [43] studied the effect of the calcination temperature on the adsorptive capacity of the $Mg/Al\text{-}CO_3$ LDH. They found out that the best properties were obtained at calcination temperature of 400 °C, which they attributed to the obtaining of a combination of surface area and the availability of the active basic sites. Actually, at this temperature the material is still amorphous, which allows having a relatively high surface area. Therefore, there is a high number of exposed basic sites, allowing the reversible CO_2 adsorption according to the following reaction:

$$Mg\text{-}O + CO_2 \;\rightarrow\; Mg\text{-}O \cdots CO_2 \,(ad)$$

(4)

However, if the LDH is calcined under 500 °C, the material is able to transform back to the original LDH structure when it is exposed to a carbonate solution or another anionic containing solution. Finally, if the sample is heated to temperatures above 500 °C, the structural changes become irreversible because of the spinel phase formation [37].

As mentioned, the mixed oxides derived from the LDH calcination possess some interesting characteristics such as high specific surface area, excess of positive charge that needs to be compensated, basic sites and thermal stability at elevated temperatures (200 – 400 °C). Besides these aspects, it is important to consider the advantage of acid-base interactions on the CO_2 sorption applications, where acidic CO_2 molecules interact with the basic sites on the derived oxide. These characteristics make the LDH-materials acceptable CO_2 captors [43, 45]. However, the CO_2 adsorption capacity of this material is low compared with other ceramic sorbents; reaching mean values smaller than 0.1 mmol/g [46]. Nevertheless, many studies suggest that the adsorption capacity of LDH materials can be improved by modifying a factor set such as: composition, improvement of the material's basicity and contaminant gas stream composition [30-32, 36, 41-45, 47-59].

As previously mentioned, Reddy et al. [43] studied the influence of the calcination temperature of LDHs on their CO_2 capture properties. The Mg_3/Al_1-CO_3 material was calcined at different temperatures ranging from 200 to 600 °C. The results showed that when the calcination temperatures are under 400 °C, LDH is considered to be dehydrated and materials still keep the layered structure intact, wherein the CO_3^{2-} ions are occupying the basic sites. The obtained samples calcined at 400 °C have the maximum BET surface area of 167 m^2/g compared with samples calcined at lower temperatures. Moreover, during the calcination of the LDH at higher temperatures (T > 500 °C), most of the CO_3^{2-} decompose to release some basic sites for CO_2 adsorption. However, the final amount of basic sites decreases with the subsequent crystallization of the MgO and spinel ($MgAl_2O_4$). Hence, LDH materials obtained at 400 °C have the highest surface area and the maximum quantities of active basic sites exposed. Because of these characteristics, they achieved a total sorption capacity of 0.5 mmol/g [43]. The same researchers observed that 88% of the captured gas can be desorbed and during the material regeneration 98% of the original weight is gained. This is another important property of LDH materials in high temperature CO_2 separation applications as described later..

As mentioned, the thermal evolution of the layered structure has a great influence on the CO_2 capture. The loss of superficial interlayer water occurs at 200 °C. Then at temperatures between 300 and 400 °C the layer decomposition begins, resulting in an amorphous 3D network with the highest surface area [30], so the adsorption temperature improves the CO_2 capture in the order of 400 > 300 > 20 >200 °C [41-42, 47, 52]

Several researchers have investigated a set of different factors to improve the CO_2 sorption capacity. Yong et al. [47, 48] studied the factors which influence the CO_2 capture in LDH materials, such as aluminum content, water content and heat treatment temperature. Regarding the M/Al-CO_3 LDHs (M = Mg, Ni, Co, Cu or Zn), the best CO_2 sorption capacity was obtained for the Mg/Al materials degassed at 400 °C and with adsorption conditions of 25 °C. In general, the sorption capacity follows the trend Ni > Mg > Co > Cu = Zn. However, when the degassed temperature is increased, the trend is modified to Mg (400 °C) > Co (300°C) > Ni (350°C) > Cu (300°C) >Zn (200°C). These results show that Mg/Al-CO_3 is the best composition at the degassing temperature of 400 °C [47]. At this temperature, the material consists of an amorphous phase with optimal properties for use as CO_2captor [42]. Also, the influence of Al^{+3} has been studied as a trivalent cation at 25 and 300 °C adsorption temperatures, by Yong [41] and Yamamoto [49] respectively. Both samples were degassed at 300 °C and the results showed that the CO_2 capture is influenced by the adsorption temperature. At a temperature of 25 °C, the maximum adsorption was 0.41 mmol/g with an Mg/Al ratio equal to 1.5 (x = 0.375) [41] and at 300 °C the amount of CO_2 adsorbed was 1.5 mmol/g for a cation ratio of 1.66 (x = 0.4) [49]. The differences between the two capacities can be attributed to the Al content differences. The Al incorporation in the structure has two functions: 1) to increase the charge density on the brucite-like sheet; and 2) to reduce the interlaminar distance and the number of sites with high resistance to CO_2 adsorption [48].

On the other hand, Qian et al. [50] studied the effect of the charge compensation anions ($A^- = CO_3^{-2}$, NO_3^{-1}, Cl^-, SO_4^{-2} and HCO_3^{-1}) on the structural properties and CO_2 adsorption capacity of Mg/Al-A^- (molar ratio equal to 3). Despite all of the prepared LDH materials showed the typical XRD patterns of LDH materials, slight structural and microstructural differences were observed. In fact, the interlayer distance changed by varying the interlayer anions due to their difference in sizes and carried charges. These differences affect the morphology and the BET surface area of both fresh and heat-treated LDH materials. Additionally, thermal treatments were performed in order to optimize the adsorption capacity of these materials. The optimal temperature treatment was established for each Mg/Al-A^-based on the surface area of each calcined LDH. Then the CO_2 adsorption capacities of calcined LDH were tested at 200 °C. $Mg_3/Al_1–CO_3$ showed the highest CO_2 adsorption capacity (0.53 mmol/g). This value was much higher than those obtained for calcined $Mg_3/Al_1-NO_3 > Mg_3/Al_1-HCO_3$, Mg_3Al_1-Cl, and Mg_3/Al_1-SO_4 (≈ 0.1 mmol/g). The results indicated that BET surface area of calcined LDHs seems be the main parameter that determines the CO_2 adsorption capacity because the Mg-O active basic site [43, 45].

It has been demonstrated that the quasi-amorphous phase obtained by the thermal treatment of LDH at the lowest possible temperature has the highest CO_2 capture capacity. This finding is in line with the fact that high calcination temperature can decrease the number of active Mg–O sites due to the formation of crystal MgO [51].

Yong [41] and Yamamoto [49] investigated the influence of the several types of anions. The results suggested that the amounts CO_2 capture decrease as a function of the anion size, which promotes a larger interlayer spacing and the higher charge: $Fe(CN)_6^{4-}$(1.5 mmol/g) $> CO_3^{2-}$ (0.5 mmol/g) $> NO_3^-$(0.4 mmol/g) $> OH^-$ (0.4-0.25 mmol/g). The reason is that $Fe(CN)_6^{4-}$ and CO_3^{2-}, because they have more void space in the interlayer due size, and are able to accommodate higher CO_2 quantities. Calcined layered double hydroxide derivatives have shown great potential for high temperature CO_2separation from flue gases. However, the presence of SOx and H_2O from flue gases can strongly affect CO_2 adsorption capacity and regeneration of hydrotalcite-like compounds. Flue gases emitted from power stations contain considerable amounts of water in the form of steam. The percentage of water found in the flue gas emitted from different sources varies between 7 and 22%, with the emissions from brown coal combustion having the highest water content [45]. For many other gas adsorption sorbents, steam generally has a negative effect on the adsorption performance because of competition for basic sites between CO_2 and H_2O. However, the presence of water or steam seems to be favorable for the adsorption capacity onto LDH [31,43,53,54]. This fact is the result of the increasing potential energy that is able to further activate basic sites, possibly by maintaining the hydroxyl concentration of the surface material and/or preventing site poisoning through carbonate or coke deposition [31]. An example of the above was reported by Yong et al. [47], who found that water or steam can increase the adsorption capacity of CO_2 by about 25%, from 0.4 mmol/g to 0.5 mmol/g.

Ding et al. [31] studied CO_2 adsorption at higher temperatures (480 °C) under conditions for steam reforming of methane. They found an adsorption capacity of 0.58 mmol/g, which was independent of water vapor content in the feed. In turn, Reddy et al. [45] investigated calcined LDHs' sorption performance influenced by CO_2 wet-gas streams. LDH samples were calcined at 400 °C [43] before measuring CO_2 sorption at 200 °C. The gas streams used were CO_2, CO_2 + H_2O, flue gas (14% CO_2, 4% O_2 and 82 % N_2) +12% H_2O.

For a pure CO_2 dry sorption, the maximum weight gain was 2.72% (~0.61 mmol/g) after 60 min, whereas the wet adsorption increased the weight of the calcined LDHs to 4.81%, showing an additional 2.09%, where He and He + H_2O were used to remove the H_2O water capture. The results showed that the helium has virtually no significant sorption affinity for the material, whereas the water-sorption profile of it clearly indicates a water weight gain of 1.67%, i.e., the gain was 0.1mmol/g due to steam presence, showing that water has a positive effect, shifting the CO_2 sorption by 0.42% as compared to dry CO_2 sorption. Also, these results revealed that in all cases about 70% sorption occurs during the first 5 min and reaches equilibrium after around 30 min.

To determine the influence of CO_2, Reddy et al. [43] tested a sample in both, wet and dry CO_2 stream conditions. The experiments showed that the same quantity of CO_2 can be trapped for the solid sorbent after two hours. The presence of water in the stream only affects the kinetics of the process. This result is in agreement with that reported by Ding et al. [31]. On the other hand, the results of the material tested suggest that the fact the CO_2 capture from flue gas was higher than in a pure stream of CO_2 might have been because the polluted gas was diluted in the stream. The presence of the water does not enhance de CO_2 capture; the maximum CO_2 adsorbed was 0.9 mmol/g. The differences between Reddy et al. results and the previously mentioned studies can be caused by the use of uncalcined LDHs, which already contain an -OH network.

To apply these materials in industrial processes, it is important to know the times during which each sorbent material can be used. Tests of the cyclability in LDH materials disclose that as function of the temperature the CO_2 capture time can vary. This can be attributed to CO_2 chemisorbed during each cycle [54] and/or to the formation of spinel-based aluminas, such as -Al_2O_3 (at temperatures higher than

400 °C). Hibino et al. [52] found that the carbonate content, acting as charge-compensating anion, continuously decreases in subsequent calcination – rehydration cycles. Reddy et al. tested LDH materials during six CO_2 adsorption (200 °C)-desorption (300 °C) cycles. The average amount gained was 0.58 mmol/g, whereas 75% of this value is desorbed, reaching desorption equilibrium after the third cycle. This can be attributed to the stabilization of the material phase and basic sites during the temperature swing.

Hufton et al. [54] studied a LDH material during several cycles in dry and wet CO_2 flows. As previously discussed, the presence of steam in the flow gas improves the CO_2 adsorption. However, after 10 adsorption cycles, the capture decreased 45%. The same behavior was observed in the dry gas flow. However, the final capture was similar to the wet gas stream, in agreement with Reddy et al. [43].

Recent studies have demonstrated that K-impregnated LDH or K-impregnated mixed oxides have a better CO_2 capture capacity due to the addition of K alkaline-earth element that improves the chemical affinity between the acidic CO_2 and alkaline surface of the sorbent material [32, 36, 55-56]. Additionally, it has been proposed that K-impregnation reduces the CO_2 diffusion resistance in the material. [57]. Hufton et al. [58] showed that the K-impregnation increases the CO_2 capture, but there is an optimal quantity of K to reach the maximum capture. Qiang et al. [50] tested an Mg_3/Al_1-CO_3(pH = 10) impregnated with 20 wt.% K_2CO_3. The CO_2 adsorption capacity was increased between 0.81 and 0.85 mmol/g in the temperature range of 300 - 350 °C. This adsorption capacity is adequate for application in water gas shift reactions (WGS).

Lee et al. [59] tested the behavior of three commercial LDHs impregnated with K (K_2CO_3/LDH ratio between 0 and 1). Three Mg/Al-CO_3 LDH with different contents of magnesium were used. Results indicated that the sorption capacity of the LDH is improved by about 10 times with the optimal K_2CO_3 additions. Additionally, it was observed that impregnation is not the only factor that influences the adsorption but the composition too. The best value was obtained when the content of divalent cation was reduced and therefore, the material had a composition with the maximum trivalent cation content. The CO_2 adsorption capacity improved from 0.1mmol/g to 0.95mmol/g with K_2CO_3/LDH weight ratio equal to 0.35 at 400 °C. After determining the optimal alkaline source/LDH ratio, a set of samples was evaluated as a function of the temperature and the results showed a maximum of 1.35 mmol/g, at 50 °C. In the impregnated materials, CO_2 chemisorption can occur and the sorbed CO_2 can be further stored as metal carbonate forms.

Other alkaline elements can be used to improve the sorption capacity of materials. Martunus et al. [46] studied the impregnation of LDH with Na and K. The LDH samples were thermally treated at 450 °C for 5 min then calcined samples were re-crystallized in K_2CO_3 and Na_2CO_3 (1 M) solutions. The re-crystallized materials were tested as CO_2 captors and the capture was maximum with LDH-Na (0.688 mmol/g) > LDH- K (0.575 mmol/g) at 350 °C after five cycles. Finally, the re-crystallized material with the highest capture was calcined at 650 °C for 4 h and re-crystallized with a solution containing the appropriate quantities of K and Na to achieve alkaline metal loading up to 20%. When the sample was Impregnated with additional K and Na at 18.4% and 1.6%, respectively, the adsorption capacity rose from 0.688 to 1.21 mmol/g. This capacity increase was achievable despite the relatively low BET surface area, equal to 124 m²/g.

Other alkaline elements such as cesium have been studied as reinforcement. Oliveira et al. [55] tested commercial Mg_1/Al_1-CO_3 and Mg_6Al_1-CO_3 impregnated with K and Cs carbonates. The materials were evaluated in the presence of steam (26% v/v water content) gas at different temperatures (306, 403 and 510 °C) at 0.4 bar of CO_2 partial pressure (total pressure 2 bar). The LDH with the highest sorption capacity was Mg_1/Al_1-CO_3–K with 0.76 mmol/g at 403 °C. Among the Cs impregnated samples, the Mg_6Al_1-CO_3-Cs presented the highest capacity with 0.41 mmol/g, while the commercial LDH samples presented CO_2 sorption capacities around 0.1 mmol/g.

The results suggest the existence of a sorption mechanism combining physical adsorption and chemical reaction. First the maximum physical adsorption is reached, then the chemisorption begins, but there is an optimal temperature. If the temperature is too low, the chemisorption does not happen, but with higher temperatures the loss of porosity impedes the contact of CO_2 molecules with active basic sites promoted by the alkaline element addition.

These results suggest there is an optimum amount of K_2CO_3 to impregnate the LDH that achieves a balance between the increase in the basicity of the sorbent material and its reduction of surface area, associated with CO_2 capture capacity. The influence of potassium is currently not clear and the relevant research is still ongoing. Finally, CO_2 adsorption capacity on the synthesized 20 wt.%K_2CO_3/Mg_3/Al_1–CO_3 (pH = 10) probably could be further increased in the presence of steam.

CERAMIC OXIDE MEMBRANES AS AN ALTERNATIVE FOR CO_2 SEPARATION

Membrane-based processes, related to gas separation and purification, have achieved an important level of development for a variety of industrial applications [60]. Therefore, the use of separation membranes is one of the promising technologies for reducing the emissions of greenhouse gases such as CO_2. The term membrane is defined as a permselective barrier between two phases, the feed or upstream and permeate or downstream side [61]. This permselective barrier has the property to control the rate of transport of different species from the upstream to the downstream side, causing the concentration or purification of one of the species present in the feed gas mixture.

Membrane-based processes offer the advantage of large scale application to separate CO_2 from a gas mixture. Figure 2 schematizes the process where concentrated CO_2 is selectively separated from flue gas that is mainly composed of nitrogen and carbon dioxide along with other gases such as water vapor, SO_x, NO_x and methane. Subsequent to the membrane process, concentrated CO_2 obtained at the permeate side can be disposed or used as raw material for the synthesis of several chemicals such as fuel and value-added products [62].

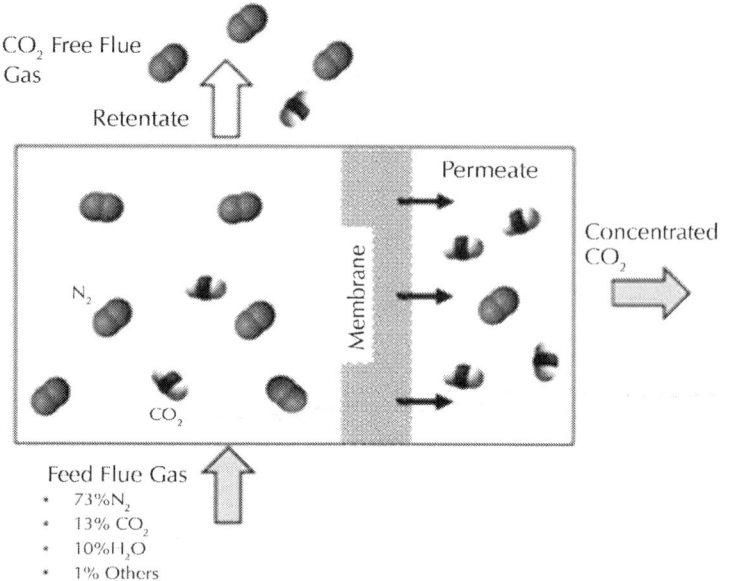

Figure 2: Membrane-based processes for the carbon dioxide separation from flue gases. The concentrated CO_2 is obtained in the permeate side.

Of course, the rate of transport or permeation properties of a particular gas through a given membrane depend on the nature of the permeant gas, as well as the physical and chemical properties of the membrane.

Inorganic membranes are more thermally and chemically stable and have better mechanical properties than organic polymer membranes; ceramic membranes offer both the advantage of large scale application and potential for pre- and post-combustion CO_2 separation applications, where membranes systems would be operating at elevated temperatures of 300-1000 °C [63].

Inorganic ceramic membranes can be classified as porous and nonporous or dense. These differ from each other not only in their structures but also in the mechanism of permeation. In porous membranes, the transport of species is explained with the pore-flow model, in which permeants are transported by pressure-driven convective flow through the pore network. Separation occurs because one of the permeants is excluded (molecular filtration or sieving) from the pores in the membrane and remains in the retentate while the other permeants move towards the downstream side. On the other hand, in nonporous membranes, separation occurs by solution-diffusion, in which permeants dissolve in the membrane material and then diffuse through the bulk membrane by a concentration gradient [60].

Porous Membranes Based on Alkaline and Alkaline-Earth Ceramic Oxides for Co_2 Separation

Among the porous systems for CO_2 separation, both microporous (carbon, silica and zeolite membranes) and modified mesoporous membranes have been reported [63-64].

Zeolites are hydrous crystalline aluminosilicates that exhibit an intracrystalline microporous structure as a result of the particular three-dimensional arrangement of their TO_4 tetrahedral units (T=Si or Al) [65]. Zeolite membranes are commonly prepared as thin films grown on porous alumina supports via hydrothermal synthesis and dry gel conversion methods [66]. Zeolite membranes of different structures have been developed to separate CO_2 from other gases via molecular sieving [67-69]. For example, membranes prepared with the 12-member ring faujasite (FAU)-type zeolite show high separation factors of 20-100 for binary gas mixtures of CO_2/N_2 [69]. In the same sense, T zeolite membranes exhibited very high selectivity, of about 400, for CO_2/CH_4 and 104 for CO_2/N_2. The high selectivity of CO_2/CH_4 exhibited by T zeolites is due to the small pore size of about 0.41 nm, which is similar in size to the CH_4 molecule but larger than CO_2 [69]. Table 1 shows the kinetic diameter of various molecules that are present in CO_2 containing gas mixtures such as flue and natural gas [70].

Table 1: Kinetic diameter of various molecules based on the Lennard Jhones relationship

Molecule	Kinetic diameter (Å)
H2O	2.65
H2	2.69
CO2	3.3

O2	3.46
N2	3.64
CH4	3.80

Deca-dodecasil 3R (DDR) (0.36 nm x 0.44 nm), and pseudo-zeolite materials like silicoaluminophosphate (SAPO)-34 (0.38 nm) also show high CO_2/CH_4 selectivities due to narrow molecular sieving, which controls molecular transport into this material [69, 71-73]. For example, Tomita et al. [74] obtained a CO_2/CH_4 separation factor of 220 and CO_2 permeance values of 7×10^{-8} mol m^{-2} s^{-1} Pa^{-1} at 28 °C on a DDR membrane [75].

As discussed, one of the most important factors controlling permeation through microporous membranes is the restriction imposed by the molecular size of the permeant. However, the transport mechanism in microporous systems is more complex than just size exclusion and the permeation and selectivity properties are also affected by competitive adsorption among perment species that produce differences in mobility [76].

Thus, the diffusion mechanism for gas permeation through microporous membranes can be characterized by two modes: one controlled by adsorption and a second one where diffusion dominates [63]. In the case of adsorption-controlled mode with permeating gases having strong affinity with the membrane, a gas permeation flux equation is obtained by assuming steady-state single gas permeation, a constant diffusivity and a single gas adsorption described by a Langmuir-type adsorption isotherm, as in Eq. (5).

$$J = \phi q_s \frac{D_c}{L}\left(\frac{1 + bP_f}{1 + bP_p}\right) \ or \ J = \phi q_s \frac{D_c}{L}\left(\frac{1 - \theta_p}{1 + \theta_f}\right)$$

(5)

where J is the permeation flux, ϕ is a geometric correction factor that involves both membrane porosity and tortuosity, D_c is the corrected diffusivity of the permeating species, L is the membrane thickness, P_f and P_p represent the feed and permeate pressure respectively and θ_f and θ_p represent the relative occupancies.

Furthermore, if the adsorption isotherm of the permeating gas is linear (1 >> bP), then flux permeation is described by Eq. (6).

$$F = \phi q_s \frac{D_c}{L}\left(\frac{D_c}{L}\right)K$$

(6)

where F is the permeance and $K = q_s b$ is the adsorption equilibrium constant. Therefore, from Eq. (5)it can be concluded that permeance is determined by both diffusivity (Dc) and adsorption (K). Based on the above, an interesting option to enhance membrane properties is to intercalate zeolite membranes with alkaline and alkaline-earth cations. Zeolite intercalation can enhance the separation between CO_2 and other molecules such as N_2 by promoting preferential CO_2 adsorption [63, 77]. It is well known that zeolites show affinity for polar molecules, like CO_2, due to the strong interaction of their quadrupole moment with the electric field of the zeolite framework. In this sense, the adsorption properties of zeolites can be enhanced by the inclusion of exchangeable cations within the cavities of zeolites where the adsorbent-adsorbate interactions are influenced by the basicity and electric field of the adsorbent cavities [78-80]. Lara-Medina et al. [77] carried out separation studies of CO_2 and N_2 with a silicalite-1 zeolite membrane prepared via hydrothermal synthesis and subsequently modified by using lithium solutions in order to promote preferential CO_2 adsorption and diffusion. CO_2/N_2 separation factor increases from 1.46 up to 6 at 25 psi and 400 °C after lithium modification. An et al. [79] studied a series of membranes prepared starting from natural Clinoptilolite zeolite rocks. Disk membranes were obtained by cutting and polishing of the original minerals, which were subsequently chemically treated with aqueous solutions containing Li, Na, Sr or Ba ions. Ionic exchanged membranes showed better permeation properties due to the presence of the extra framework cations.

Although zeolite membranes offer certain advantages in comparison with polymer membranes, such as chemical stability, the main issues are related to the selectivity decrease as a function of the permeation temperature. This is explained in terms of the contribution of the adsorption to the separation, which decreases sharply as temperature increases. At high temperature, physical adsorption becomes negligible and permeation is mainly controlled by diffusion [63, 76]. Additionally, due to the fact that CO_2 and N_2 molecules have similar sizes (Table 1), the difference in diffusivity is not a strong controlling factor in determining selectivity.

Modified -Al_2O_3 mesoporous membranes have been also reported as a means for CO_2 separation [64]. Transport mechanisms in porous membranes have the contribution of different regimes. An overview of the different mechanisms is given in Table 2.

Table 2: Transport mechanisms in porous membranes

Transport Type	Pore diameter	Characteristics
Viscous flow	>20 µm	Non selective.
Molecular diffusion	>10 µm	Affects the total flow resistance of the membrane system.
Knudsen diffusion	2 – 100 nm	Occurs when the mean free path of the molecule is much larger than pore radius of the membrane. Shows selectivity based on molecular weights.
Surface diffusion		Shows selectivity due to interaction of molecules with membrane walls.
Capillary condensation		
Micropore diffusion (Configurational diffusion)	< 1.5 nm	

Depending on the particular system, permeability of a membrane can involve several transport mechanisms that take place simultaneously. Considering no membrane defects and pore sizes in the range of 2.5-5 nm, -Al_2O_3 based membranes theoretically have two transport regimes: Knudsen diffusion and surface diffusion. Eq. (7) describes the permeability of a membrane by taking into consideration the Knudsen and surface diffusion.

$$F = \left(\frac{2\varepsilon\mu r}{3RTL}\right)\left(\frac{8RT}{\pi M}\right)^{0.5} + \frac{2\varepsilon\mu D_s}{r A_0 N_{av}}\frac{d x_s}{dP}$$

(7)

where r is the mean pore radius, μ is a shape factor, R is the universal gas constant, T is the temperature, P is the mean pressure, M is molar mass of the gas, A_o is the surface area occupied by a molecule, D_s is the surface diffusion coefficient, N_{av} is Avogadro's constant and X_s is the percentage of occupied surface compared with a monolayer [81].

For the cases when Knudsen diffusion dominates, selectivity can be correlated to the molecular weights of the permeating gases by the so called Graham's law of diffusion, which establishes that the transport rate of any gas is inversely proportional to the square root of its molecular weight. The CO_2/N_2 separation factor considering pure Knudsen diffusion is given by Eq. (8) and has a value of just 0.8. Therefore, Eq. (8) clearly shows that separation via Knudsen is limited for systems where species are of similar molecular weight.

$$\alpha\left(\frac{CO_2}{N_2}\right) = \sqrt{\frac{M_{CO_2}}{M_{N_2}}}$$

(8)

Based on the aforesaid, CO_2/N_2 separation factor can be better enhanced by promoting the surface diffusion mechanism (second term on the right hand side of Eq. (7)). Surface diffusion involves the adsorption of gas molecules on the surface of the pore and subsequent diffusion of the adsorbed species along the surface by a concentration gradient. Then separation properties of a membrane can be improved by generating such an interaction between one component of the feed gas mixture with the membrane; one option being via a chemical modification.

Cho et al [81] prepared a series of thin (2-5 µm thickness) -Al_2O_3 and CaO- or SiO_2-modified -Al_2O_3 membranes for CO_2 separation at temperatures between 25 and 400 °C. Impregnation of membranes with SiO_2 or alkaline CaO was done in order to improve the CO_2/N_2

selectivity by promoting adsorption between CO_2 gas molecules and the membrane pore wall. Although this kind of chemical modification of the membrane surface and the pore walls is able to activate the surface diffusion mechanism, an interesting behavior was observed. The CO_2/N_2 separation factor increased from 1.0 to 1.38 at 25 °C after modification of the -Al_2O_3 with SiO_2. On the other hand, CaO impregnated membranes showed a separation factor of 0.98, which is even lower than that of the unmodified -Al_2O_3. The same behavior has been reported by Uhlhorn et al. [82-83]. They reported MgO modified -Al_2O_3 membranes which did not show significant enhancement in the permeation and selectivity properties as a result of the modification process. This fact was explained in terms of the surface diffusion mechanisms. As discussed, it is expected that physicochemical modifications of the membrane can enhance preferential adsorption of the gas species in the feed. Impregnations with alkaline oxide such as calcium oxide or magnesia on the -alumina surface give more strong base sites than those promoted by silica. Therefore, it promotes a strong bonding of CO_2 on the alumina surface, causing CO_2 molecules to lose mobility, resulting in a smaller contribution of surface diffusion to the total transport mechanism.

There is another kind of membrane where alkaline and alkaline-earth ceramic oxides have been used for the fabrication of CO_2 permselective membranes. In these cases ceramic materials were chosen because of their well-known properties of physisorption of CO_2 at low and intermediated temperatures.

Kusakabe et al. [84] prepared both pure and modified $BaTiO_3$ CO_2 permselective membranes via the alkoxide based sol-gel method; impregnation and calcination at 600 °C. In order to establish the effects of CO_2 partial pressure, temperature and influence of the secondary oxide presence (CuO, MgO or La_2O) on the CO_2 adsorption properties of the membranes, pure and modified barium titanate powders were first evaluated by thermogravimetry and chromatography techniques. Dynamic CO_2 absorption was evaluated by applying the impulse response method, wherein the $BaTiO_3$ powder was packed in a separation column. The results suggested that the CO_2 molecules adsorbed on the $BaTiO_3$ powder are mobile at temperatures about 500 °C. Therefore, this membrane exhibits CO_2 permeation due to surface diffusion mechanism. Even though the prepared membranes showed selectivity, the Knudsen diffusion still has an important contribution to the gas transport due to the presence of membrane defects. The maximum separation factor of CO_2/N_2 through the membranes was estimated as 1.2. Therefore, further improvement of the permeation properties of this kind of membrane requires obtaining pinhole-free membranes.

Based on the same criteria, Nomura et al. [85] prepared Li_4SiO_4-based thin membranes on porous alumina supports. Membranes were obtained by the thermal treatment of different silica containing porous materials (Silicalite-1 and mesoporous silica) impregnated with lithium compounds. The authors called this method solid conversion. The use of different silica porous sources was proposed in order to enhance the reaction rate of Si and Li on the porous support at relatively low temperature, avoiding the reaction between the Li and alumina support itself. In the case of Silicalite-1 (MFI zeolite), a zeolite thin film was first prepared on the support by following the dry gel conversion technique. Then, the prepared Silicalite-1 layer was impregnated via dipping into a slurry containing lithium and silica fumed reactants (Li:Si = 4:1) and subsequently into a Li_2CO_3-K_2CO_3 slurry. The membrane was finally calcined at 600 °C for 2 h. It is believed that carbonate melts to fill the cracks and the pinholes of the Li_4SiO_4 formed membrane. A similar procedure of coating and calcination was carried out to prepare high quality membranes starting from mesoporous silica sources with pore sizes of 1.8-12.8 nm. Precursors react to form a Li_4SiO_4 membrane of 2-5 µm thickness that exhibits an N_2 permeance of 1.8×10^{-9} mol m^{-2} s^{-1} Pa^{-1} at 400 °C. This suggests there are no big defects after impregnation of the membrane with the binary mixture of Li_2CO_3-K_2CO_3 carbonate. Due to the fact that the membrane operates in a rich CO_2 atmosphere, carbonates do not decompose even at temperatures of 600 °C. The maximum CO_2/N_2 permeance ratio was 0.85. The separation factor was higher than that for the Knudsen diffusion. Therefore, it can be conclude that Li_4SiO_4 layer was selective to CO_2 over N_2 at high temperature of 600 °C.

Nomura [86] reported a two -stage approach for the preparation of Li_4SiO_4-CO_2 selective membranes that involves the fabrication of a supported Li_4SiO_4 membrane and its subsequent modification by using a chemical vapor deposition (CVD) method. First, for the preparation of a thin Li_4SiO_4 membrane the so called solid conversion method described before was used, which is based on the reaction between a porous silica source and a lithium containing solution coated on a porous alumina membrane support. Although the formed membranes showed certain selectivity due to the preferential adsorption of CO_2 over N_2, the presence of pinholes and cracks caused low separation factors. Therefore, the membrane defects were fixed by using the counter diffusion CVD method to form a silica coating that fills the gaps between the lithium orthosilicate particles that make up the membrane. N_2 permeance was reduced about three orders of magnitude after CVD modification. Nitrogen permeance before and after the CVD treatment was 3.4×10^{-6} mol m^{-2} s^{-1} Pa^{-1} and 1.2×10^{-9} mol m^{-2}s^{-1} Pa^{-1} respectively. In the same sense, the CO_2/N_2 permeance rate increased from 0.7 to 1.2 at 600 °C. Some issues related with this system are the chemical and structural stability of the membranes observed during the permeation tests at elevated temperature. The membranes were broken when permeation tests were carried out at temperatures higher than 700 °C, with the consequent decrease in the CO_2/N_2 selectivity. The aforesaid is the result of the CO_2 chemisorption on the membrane. Lithium orthosilicate reacts with CO_2 to form lithium carbonate and lithium metasilicate (Li_2SiO_3) as products, as indicated by Eq. (9).

$$Li_4SiO_4 + CO_2 \longleftrightarrow Li_2CO_3 + Li_2SiO_3$$

(9)

Thermodynamically, this reaction is prone to occur at temperatures between room temperature and about 700 °C. However, experimentally it has been observed that reaction kinetics sharply increase above 550 °C. At these temperatures, the formation of carbonates involves an important change in volume that ends in the membrane's rupture.

Therefore, one of the issues related to the development of this kind of inorganic membrane is the thermochemical stability. Due to reactivity of alkaline and alkaline-earth ceramic oxides with CO_2 to form carbonates, not only preferential adsorption of CO_2 molecules over N_2 occurs, but CO_2 chemisorption and reaction.. Therefore, it is mandatory to establish the operational temperature within a range where CO_2 selective adsorption on the membrane layer promotes the separation process without reaction.

Nonporous Membranes Based on Alkaline and Alkaline-Earth Ceramic Oxides for Co_2 Separation

Some researchers have proposed the use of alkaline and alkaline-earth ceramic oxides to prepare membranes that are able to separate CO_2 at high temperatures via a different transport mechanism than those observed on porous membranes. Li_2ZrO_3 and Li_4SiO_4 based membranes are examples of the aforesaid. Permselectivity of CO_2 through these membranes takes place not only due to the selective CO_2 adsorption properties of ceramic phases but also via a mechanism of gas separation that involves the transport of CO_3^{2-} and O^{2-} ionic species through the electrolytes (carbonate-metal oxide) phases formed by the reaction of the membrane with the CO_2 [87-89].

Kawamura et al. [87] fabricated and characterized a membrane for CO_2 separation at high temperatures. The membrane was made of lithium zirconate (Li_2ZrO_3), an alkaline ceramic oxide that reacts with CO_2 to produced Li_2CO_3 and ZrO_2. These two reaction products are electrolyte materials produced *in-situ* when the membrane is exposed to the rich carbon dioxide atmosphere. The electrolytes formed thus are capable to transport both CO_2 and O_2 across the membrane via a dual ion conduction mechanism. The prepared membrane exhibited a separation factor of 4.9 between CO_2 and CH_4 gas molecules at a temperature of 600 °C. The obtained separation factor is higher than the Knudsen diffusion limit, 0.6. Therefore, the results clearly suggest the potential use of this kind of membrane system for CO_2 separation such as the case of CO_2 removal from natural gas.

Yamaguchi et al. [88] investigated the concept of the dual-ion conduction facilitated mechanism previously observed for the case of Li_2ZrO_3 membranes by focusing their efforts on the preparation of a CO_2 permselective membrane based on lithium orthosilicate (Li_4SiO_4). The supported membrane was prepared via a dip coating technique by using Li_4SiO_4 suspensions. The coating process was repeated several times before impregnation of the membrane with a Li_2CO_3/K_2CO_3 carbonate mixture and final sintering at 750 °C. In this membrane system, Li_4SiO_4 reacts *in-situ* with CO_2 to form Li_2CO_3 and Li_2SiO_3.

Gas separation studies were performed by using CO_2/N_2 mixtures as feed gas. The observed CO_2 permeance values were about 1×10^{-8} mol $m^{-2}s^{-1}Pa^{-1}$ in the temperature range of 525-625 °C. The CO_2/N_2 separation factor was estimated between four and six. Figure 3 shows a scheme of the dual-ion conduction mechanism explained as follows. In the feed side, carbon dioxide dissolves in the material and diffuses as carbonate ions through the molten carbonate electrolyte due to a concentration gradient. Then, in the downstream side of the membrane, the formation of gaseous CO_2 implies the formation of oxygen ions which must diffuse back to the feed side across the membrane and apparently through the formed Li_2SiO_3 skeleton to obtain the charge balance.

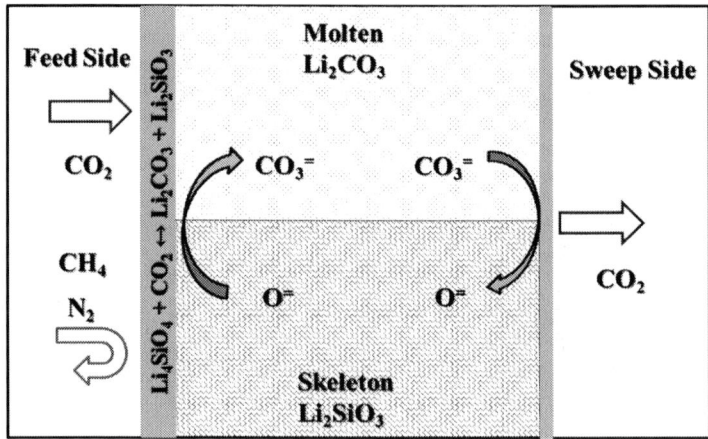

Figure 3: Schematic representation of a membrane system for the CO_2 separation via a dual-ion conduction mechanism.

The proposed transport mechanisms supports the higher selectivity values observed in the permeation test for both systems, Li_2ZrO_3 and Li_4SiO_4. Figure 4 shows the separation factor values (CO_2/N_2) obtained for different ceramic membranes described in the present report. The pure Knudsen value is written as baseline and separation factor of nonporous Li_4SiO_4 for comparison purposes. However, it is important to mention that the oxygen ion diffusion process is not totally clear. Indeed, there is no experimental study regarding the oxygen ionic conductivity properties of Li_2SiO_3 phase. On the other hand, pure ZrO_2 exhibits poor bulk oxygen ion conductivity. In fact, good conduction properties are observed only in acceptor-doped ZrO_2 based materials with oxygen vacancies being the predominant charge carriers [90]. Therefore, oxygen ion conduction through the membrane must be related to different transport paths, such as grain boundaries and interfacial regions formed between the ceramic and molten carbonate on the membrane.

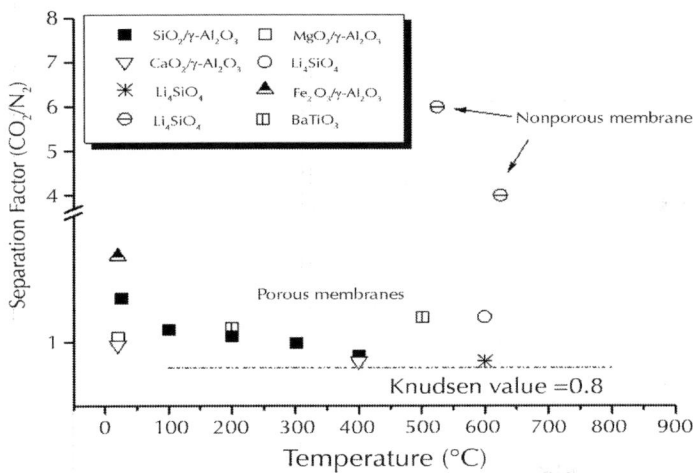

Figure 4: CO_2/N_2 separation factor of different ceramic oxide membranes.

More recently, the promising concept of ceramic oxide-carbonate dual-phase membranes has been proposed for carbon dioxide selective separation at intermediate and high temperatures (450-900 °C) [91-97].

This concept involves the fabrication of nonporous membranes capable of selectively separating CO_2 via its transport, as carbonate ions. Dual phase membranes are made of an oxygen ion conductive porous ceramic phase that hosts a molten carbonate phase. Rui et al. [98] proposed the CO_2 separation by the electrochemical conversion of CO_2 molecules to carbonate ions (CO_3^{2-}), which are subsequently transported across the membrane. Carbonate ionic species (CO_3^{2-}) are formed by the surface reaction between CO_2 and oxygen that comes from the ceramic oxide phase (feed side, Eq.(10)) and then transport of CO_3^{2-} takes place through the molten carbonate.

$$CO_2 + O_O^x \longleftrightarrow CO_3^{2-} + V_O^{\cdot\cdot}$$

(10)

Once carbonate ions have reached the permeate side, molecular CO_2 is released to the gas phase, delivering O_O^x species back to the ceramic oxide solid phase. This process takes place due to a chemical gradient of CO_2 in the system (Figure 5). Here, it is important to emphasize that dual-phase membranes are nonporous and therefore exhibit high separation selectivity as a result of the transport mechanism. Figure 5 also shows the SEM image of the cross section of a ceramic oxide-carbonate membrane prepared by pressing $La_{0.6}Sr_{0.4}Co_{0.8}Fe_{0.2}O_{3-}$ powders and subsequent infiltration of the obtained porous ceramic (bright phase) with carbonate (dark phase).

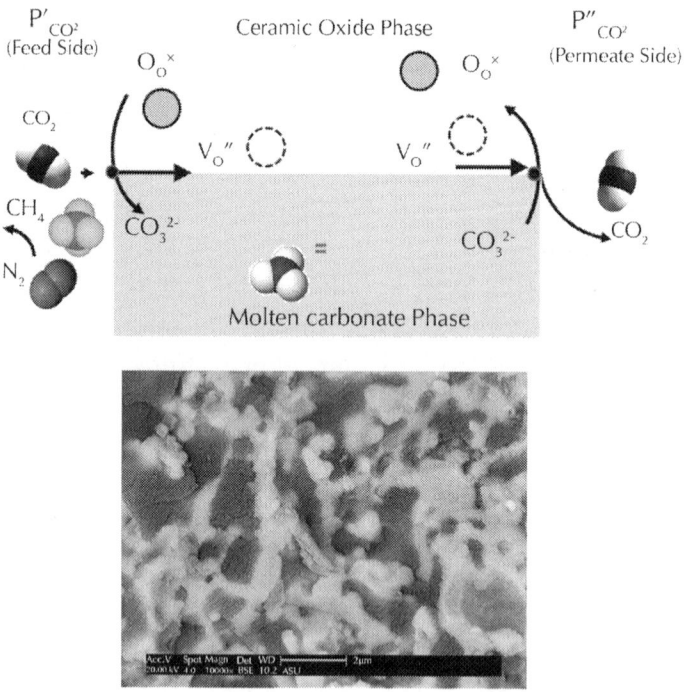

Figure 5: Schematic representation of a membrane system for the CO_2 separation and SEM image of a ceramic oxide-carbonate dual-phase membrane.

Table 3 summarizes the different studies reported and certain advances that have been achieved so far regarding the dual-phase membrane concept. This table also includes the Li_2ZrO_3 and Li_4SiO_4 nonporous membranes previously described. Although the original reports do not clearly explain the operational mechanism [26-27], the dual-phase membrane concept gives a much better idea of the possible phenomenology involved [30,33,36].

Table 3: Reported studies on dual-phase and related membranes for CO_2 separation

Ceramic Oxide phase	Molten Carbonate phase	Membrane features	Preparation method	Permeance (mol.s.-1m.-2Pa.-1)	Separation Factor (CO2/N2)	Ref.
Li2ZrO3	Li2CO3	Thick membrane	In situ by exposing Li2ZrO3 to CO2 atmosphere	1 x10-8	4.9 (CO2/ CH4 at 600°C)	[87]
Li4SiO4/ Li2SiO3	K2-Li2CO3	Thin supported membrane	Impregnation of carbonate	2x10-8	5.5 (at 525°C)	[88]
La0.6Sr0.4Co0.8Fe0.2O3-	Li-Na-K2CO3	Thick membrane (0.35-1.5 mm)	Pressing and direct infiltration	4.77 x 10-8	225 (at 900°C)	[91]
8 mol% Yttria doped zirconia (YZS) 10 mol% Gadolinia doped ceria (GDC)	Li-Na-K2CO3 Li-Na2CO3	Thin freestanding membranes (200-400 μm)	Tape casting and in situ infiltration	2.0 x10-8(YSZ) 3.0 x10-8(GDC)	> 2 (at 800 °C)	[92]
Ce0.8Sm0.2O1.9	Na2-Li2CO3	Thick membrane (1.2 mm)	Pressing of SDC-NiO powders where NiO is a sacrificial template	~1.2 x 10-6	155-255 (at 700°C)	[93]
Bi1.5Y0.3Sm0.2O3	Li-Na-K2CO3	Thin supported membrane (50 μm)	Dip coating of modified thick support and infiltration	1.1 x 10-8	2 (at 650°C)	[94]

8 mol% Yttria doped zirconia (YZS)	Li-Na-K2CO3	Thin supported membrane (10 µm)	Dip coating of YSZ on nonwettable thick support and infiltration	~ 7.8×10−8	---	[95]
Ce 0.8Sm 0.2 O1.9	Li-Na-K2CO3	Thin tubular membrane	Centrifugal casting and direct infiltration			[96]
La0.6Sr 0.4Co 0.8Fe 0.2O3-	Li-Na-K2CO3	Thick disk-shaped membrane	Pressing and direct infiltration			[97]

Applications of CO_2 Permselective Ceramic Oxide Membranes for the Design of Membrane Reactors.

As mentioned, CO_2 can be used as raw material for the synthesis of several chemicals [99]. Moreover, if CO_2 is concentrated or separated by a membrane system exhibiting high CO_2 permeation and permselectivity, this open up the possibility to develop a continuous process of membrane reaction to simultaneously capture and chemically convert CO_2. For example, if the membrane is able to separate CO_2 at intermediate and even high temperatures, it can be used for the design of a membrane reactor for the production and purification of hydrogen and syngas. Syngas is a gaseous fuel with a main chemical composition of CO, H_2, CO_2, and CH_4. Syngas can be used as feedstock for the synthesis of several other clean fuels such as H_2, methanol, ethanol, diesel and other hydrocarbons synthesized via the Fischer-Tropsch process [100-104].

Among the different processes for the synthesis of syngas and hydrogen, CO_2 methane reforming Eq. (11) and the water-gas shift reaction (WGS) Eq. (12) are the most promising options.

$$CH_4 + CO_2 \longleftrightarrow 2CO + 2H_2$$

(11)

$$CO + H_2O \longleftrightarrow CO_2 + H_2$$

(12)

Figure 6 schematizes the membrane reactor concept considering the two reactions described above.Figure 6A shows a membrane reactor for dry reforming of methane to produce syngas at temperatures between 700 and 800 °C. Figure 6B illustrates the use of ceramic oxide membranes for hydrogen purification by separating the CO_2 from water-gas shift products at about 550 °C. Additionally, Figure 6B shows the possibility of using a ceramic sorbent to chemically trap the permeate CO_2 and therefore enhance the CO_2 permeation process by reducing the concentration of CO_2 in the permeate side.

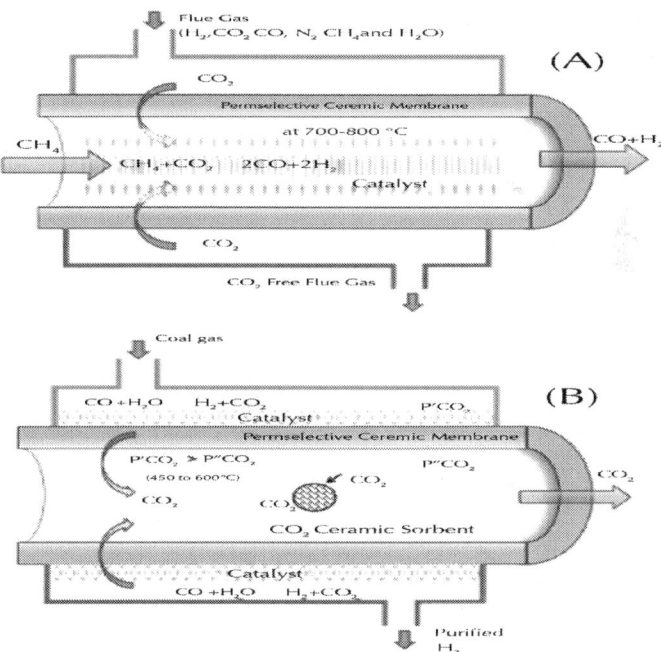

Figure 6: Schematic representation of the membrane reactor concept using a CO_2 permselective ceramic membrane: (a) CO_2 dry methane reforming and (b) water-gas shift reaction with hydrogen purification wherein CO_2 capture promotes the separation process.

CHEMICAL TRANSFORMATION OF CO_2 CATALYZED BY CERAMIC MATERIALS: THE USE OF NEW ALTERNATIVES

One of the most widely used chemical absorption techniques for carbon capture and storage/sequestration (CCS) is CO_2 adsorption by ceramic materials. Once CO_2 has been captured-fixed, it can be converted into value-added products such as precursors in chemical transformation reactions. CO_2 is extensively used for enhanced oil recovery, as a monomer feedstock for urea and polymer synthesis, in the food and beverage industry as a propellant, and in production of chemicals. Therefore, the capture-fixation of CO_2 would make a system suitable for accomplishing chemical transformation of CO_2. The utilization of carbon dioxide is also very attractive because it is environmentally benign [105-115]. CO_2 conversion to fuel and value-added products is an ideal route for CO_2 utilization due to the simultaneous disposal of CO_2 and the benefit that many products can be used as alternate transportation fuels [116]. CO_2 chemical transformation methods include (i) reverse water-gas shift, (ii) hydrogenation to hydrocarbons, alcohols, dimethyl ether and formic acid, (iii) reaction with hydrocarbons to syngas, (iv) photo- and electrochemical/catalytic conversion, and (v) thermo-chemical conversion [100-122].

CO_2 can be catalyzed to valuable organic or inorganic compounds, where some basic catalytic materials (containing alkaline or alkaline-earth elements) are used. The activation of CO_2 by alkali metals has received considerable attention in various surface science studies, which have demonstrated the formation of intermediate CO_2, dissociation of CO_2 and formation of oxalate and carbonate alkali compounds [118-121]. Carbon dioxide has been identified as one such potential vector molecule (through reduction to syngas, methanol, methane, formic acid, formaldehyde, dimethylether (DME) and short-chain olefins) [117-118, 120-122]. CO_2 is a kinetically and thermodynamically stable molecule, so CO_2 conversion reactions are endothermic and need efficient catalysts to obtain high yield. CO_2 conversion to carbon monoxide (CO) looks like the simplest route for

CO_2 reduction [121]. CO is a feedstock or intermediate product for the production of methanol and hydrocarbon fuels via Fischer-Tropsch synthesis of CH_4/CO_2 reforming to form syngas (CO/H_2) [122]. CO_2 reforming with CH_4 is an example of CO_2 being used as a soft oxidant, where the dioxide is dissociated into CO and surface oxygen, and oxygen abstracts hydrogen from methane to form water via the water-gas shift reaction (WGS) (Eqs. 11 and 12) [100-103, 121]. The catalytic chemistry of the reverse water-gas shift reaction and the following transformation to methanol/DME (or hydrocarbons via Fischer-Tropsch synthesis), and the subsequent production of gasoline (methanol-to-gasoline or diesel via hydrocracking of the alkanes produced in the Fisher-Tropsch process) are well established [102, 117-122]. On the other hand, methanol can be produced directly from carbon dioxide sources by catalytic hydrogenation and photo-assisted electrochemical reduction. A wide variety of CO_2 photo-reduction methods have been performed to oxygenate products, including formic acid (HCOOH) and formaldehyde (HCHO). HCOOH and HCHO are the simplest oxygenates produced from the reduction of CO_2 with H_2O (or proton solvents) [121]. Furthermore, CO_2 can be utilized as a monomeric building block to synthesize various value-added oxygen-rich compounds and polymers under mild conditions. As an example, chemical conversion of CO_2 through C–N bond formation can produce value-added chemicals such as oxazolidinones, quinazolines, carbamates, isocyanates and polyurethanes [105]. These commodity chemicals have been synthesized from green methods and have important applications in the pharmaceutical and plastic industries. The chemisorption of CO_2 based on C–N bond formation could be one of the most efficient strategies, utilizing liquid absorbents such as conventional aqueous amine solutions, chilled ammonia, amino-functionalized ionic liquids, and solid absorbents including amino-functionalized silica, carbon, polymers and resins. The processes by which chemicals for CO_2 capture are manufactured should also be considered in terms of their energy requirements, efficiencies, waste products, and CO_2 emissions [105, 123]. In that sense, dimethyl carbonate (DMC) is a promising target molecule derived from CO_2 catalyzed by inorganic dehydrating agents such as molecular sieves [107]. Dimethyl carbonate has received much attention as a safe, non-corrosive, and environmentally friendly building block for the production of polycarbonates and other chemicals, an additive to fuel oil owing to its high octane number and an electrolyte in lithium batteries due to its high dielectric constant. It can be synthesized through a two-step transesterification process utilizing CO_2 as raw material [105, 107].

As a complementary technology to carbon sequestration and storage (CSS), the chemical recycling of carbon dioxide to fuels is an interesting opportunity. Chemical compounds such as alkane products ($C_nH_{(2n+2)}$) are un-branched hydrocarbons suitable for diesel fuel and jet fuel [121]. In this regard, biofuels or biodiesel, catalyzed using ceramic materials, can provide a significant contribution in energy independence and mitigation of climate change [109-127]. Today the main renewable biofuels are bioethanol and biodiesel. Biodiesel is a liquid fuel consisting of mono alkyl esters (methyl or ethyl) of long-chain fatty acids derived from vegetable oils, animal fats or micro and macro algal oils [127]. Biodiesel is a sustainable, renewable, non-toxic, biodegradable diesel fuel substitute that can be employed in current diesel engines without major modification, offering an interesting alternative to petroleum-based diesel [106, 111-115, 124-128]. Besides this, it is free from sulfur and aromatic components, making it cleaner burning than petroleum diesel. Biodiesel has a high flash point, better viscosity and caloric power similar to fossil fuels. It can be mixed with petroleum fossil fuel at any weight ratio or percentage, and it can be used without blending with fossil fuel (B100) as a successful fuel [127, 128]. It has similar properties (physical and chemical) to petroleum diesel fuel. Recently, transesterification (also called alcoholysis) has been reported as the most common way to produce biodiesel with lipid feedstock (such as vegetable oil or algal oil) and alcohol (usually methanol or ethanol), in presence of an acid or base catalyst. Transesterification is the best method for producing higher-quality biodiesel and glycerol [108, 110-115, 124-132]. The reaction is facilitated with a suitable catalyst [129-131]. The catalyst presence is necessary to increase both, the reaction rate and the transesterification reaction conversion yield. The catalysts are classified as homogeneous or heterogeneous. Homogeneous catalysts act in the same liquid phase as the reaction mixture. Conversely, if the catalyst remains in a different phase, the process is called heterogeneous catalytic transesterification [113, 127-131]. Heterogeneous catalysts are mostly applied in transesterification reaction due to many advantages such as easy catalyst separation and reusability, improved selectivity, fewer process stages, no water formation or saponification reaction, including in green technology, and cost effectiveness [127, 132]. The heterogeneous catalysts increase the mass transfer rate during the transesterification reaction [127, 131]. Various ceramic materials have been investigated for the production of biodiesel [106, 109-115, 124-179]. Some of these solid catalysts include alkali and alkaline-earth metal carbonates and oxides such as magnesium oxide (MgO), calcium oxide (CaO), barium oxide (BaO), strontium oxide (SrO) [124-131, 133-143]; lithium base ceramics (Li_4SiO_4 and Li_2SiO_3 [144-146]); sodium silicate (Na_2SiO_3 [147]); transition metal oxides and derivatives (titanium oxide, zinc oxide, mixed oxides catalysts [148-149]); ion exchange resin type acid heterogeneous catalysts [150]; MCM-metal impregnated materials [114]; layered double hydroxides (hydrotalcite-like hydroxides) [151-154]; hydrocalumite-like compounds [110,155]; supported bases [156-163]; and zeolites [164-165].

Among the alkaline earth metal oxides, CaO is a promising basic heterogeneous catalyst for synthesizing biodiesel at mild temperatures (below the boiling point of methanol, MeOH) and at atmospheric pressure due to its plentiful availability and low cost, but it is rapidly hydrated and carbonated upon contact with room temperature air. CaO is the most widely used catalyst for transesterification and produces a high yield of 98% of fatty acid methyl esters (FAME) during the first cycle of reaction [130]. Granados et al. [142] used activated CaO as a solid base catalyst in the transesterification of sunflower oil to investigate the role of water and carbon dioxide on the deterioration of the catalytic performance upon contact with air for different periods. The study showed that CaO was rapidly hydrated and carbonated in air. Consequently, the reusability of the catalyst for subsequent steps is a big question mark. Di Serio et al. [170] reported a 92% biodiesel yield with MgO catalyst, using 12:1 methanol to oil molar ratio with 5.0wt% of the catalyst at methanol supercritical condition for 1 h. Wen et al. [171] carried out transesterification from waste cooking oil with methanol at 170 °C for 6 h with 10wt% of MgO/TiO_2 and 50:1 M ratio of MeOH and oil. Guo et al. [172] studied the methyl ester yield produced via transesterification of soybean oil using sodium silicate as a catalyst. Sodium silicate was an effective catalyst for the microwave-irradiated production of biodiesel and hydrothermal production of hydrogen from by-product glycerol combined with Ni catalyst. The optimum reaction conditions obtained were 7.5:1 M ratio of alcohol/oil, 3wt% catalyst amount, 1 h reaction time and 60 °C reaction temperature. The FAME yield was ~100%. On the other hand, microwave-assisted transesterification of vegetable oil with sodium silicate is an effective and economical method for the rapid production of biodiesel. The reused catalyst after transesterification process for four cycles was recovered. Overall, sodium silicate was fully used in biodiesel production and glycerol gasification, and this co-production process provided a novel green method for biodiesel production and glycerol utilization [172].

Several techniques have been investigated for the transesterification reaction using heterogeneous catalysts for biodiesel production, as follows: transesterification via radio frequency microwaves, alcohol reflex temperature, alcohol supercritical temperature and ultrasonication [127, 173-177]. Recently, the use of ultrasonic irradiation has gained interest in biodiesel production [173-177]. Ultrasonic energy can emulsify the reactants to reduce the catalyst requirement, methanol-oil ratio, reaction time and reaction temperature and also provides the mechanical energy for mixing and the required activation energy for initiating the transesterification reaction [173-176]. The ultrasound phenomenon has its own physical and chemical effects on the liquid-liquid heterogeneous reaction system through cavitation bubbles, according to the following principles [175]: (1) the chemical effect, in which radicals such as H^+ and OH^- are produced during a transient implosive collapse of bubbles (in a liquid irradiated with ultrasound), which accelerates chemical reaction in the bulk medium; and (2) the physical effect of emulsification, in which the microturbulence generated due to radial motion of bubbles leads to intimate mixing (homogenizing the mixture) of the immiscible reactants. Accordingly, the interfacial region between the oil and alcohol increases sharply, resulting in faster reaction kinetics and higher conversion of oil and biodiesel yield [127]. In 2000, the ultrasonication reactor was first introduced by Hielscher Ultrasonic GmbH for biodiesel production. Nishimura et al. [175] studied the transesterification of vegetable oil using low-frequency ultrasound (28-40kHz). An excellent yield (~98%) was obtained at a 28 kHz ultrasound while a significant reduction of reaction time was obtained by using 40 kHz ultrasound. Salamatinia et al. [176] used ultrasonic assisted transesterification to improve the reaction rate. In this study, they used SrO and BaO as heterogeneous catalysts in the production of biodiesel from palm oil. The results showed that the basic properties of the catalyst were the main cause for their high activity. The low-frequency ultrasonic assisted transesterification process had no significant mechanical effects on SrO, but BaO catalyst study confirmed that the ultrasound treatment significantly improved the process by reducing the reaction time to less than 50 min at a catalyst loading of 2.8wt% to achieve biodiesel yield higher than 95%. Another study of alkali earth metals was carried out by Mootabadi et al. [177]. They reported the effect of ultrasonic waves at 20 kHz and 200W on the regenerated catalyst and compared mechanical stirring and ultrasonic irradiation. They investigated the optimum conditions, using palm oil for biodiesel production with catalysts such as CaO, SrO and BaO. They concluded that catalyst leaching was the main cause for the catalyst inactivity in the case of the re-used catalyst. BaO catalyst was found to be stable during the leaching. At the optimized condition, 95.2% yield was achieved with 60 min of reaction time for both BaO and SrO catalysts. For CaO catalyst, 77.3% yield was achieved with the same conditions. The use of ultrasound showed great enhancement of the reaction parameters in terms of the obtained yield and reaction time. The obtained yields were 30 to 40% higher in comparison to the corresponding results obtained using a conventional stirring reactor system without ultrasonication. Deng et al. [178] prepared nano-sized mixed Mg/Al oxides. Due to their strong basicity, the nanoparticles were further used as catalyst for biodiesel production from jatropha oil. Experiments were conducted with the solid basic catalyst in an ultrasonic transesterification reaction. Under the optimum conditions, biodiesel yield was 95.2%. After removing the glycerol on the catalyst surface, the nano-sized mixed Mg/Al oxides were reused eight times. The authors concluded that calcination of hydrotalcite nanocatalyst under ultrasonic radiation is an effective method for the production of biodiesel from jatropha oil. The activity of base solid catalysts is associated to their basic strength, such that the most basic catalyst showed the highest conversion. In another work, Deng et al. [179] reported optimum conditions for biodiesel production in the presence of base solid catalysts. They studied BaO and Ca–Mg–Al hydrotalcite (the most effective). The 95% biodiesel yield from jatropha oils and Ca–Mg–Al hydrotalcite was established with 30 min of reaction time. Ca–Mg–Al hydrotalcite could be reused twelve times after washing of the adsorbed glycerol from the catalyst surface with ethanol. Other types of heterogeneous catalysts under ultrasonic irradiation were used for transesterification by Georgogianni et al. [114]. They studied a wide range of catalysts including Mg-MCM-41, Mg–Al hydrotalcite and K+-impregnated zirconium oxide. They mixed frying oils, methanol and the catalyst in a batch reactor with mechanical stirring for 24 h and with ultrasonication for 5 h. The results suggested that the basic strength was the cause of the good activity of the catalysts. Mg–Al hydrotalcite achieved the highest reaction conversion of 87% at a reaction temperature of 60 °C. Overall, ultrasonic irradiation significantly enhanced the reaction rate, causing a reduction in reaction time, and the biodiesel yield increased [114]. Consequently, a better understanding of the use of ultrasonic sound waves to accelerate the transesterification process could lead to substantial future improvement of both batch and continuous production systems, to obtain a more sustainable biodiesel production process [127].

REFERENCES

1.   Ortiz-Landeros J.; Ávalos-Rendón T. L.; Gómez-Yáñez C.; Pfeiffer H. *Analysis and Perspectives Concerning CO2 Chemisorption on Lithium Ceramics Using Thermal Analysis.* J. Therm. Anal. Calorim. 2012, 108, 647−655.

2.   Ávalos-Rendón T.; Casa-Madrid J.; Pfeiffer H. *Thermochemical Capture of Carbon Dioxide on Lithium Aluminates (LiAlO2 and Li5AlO4): A New Option for the CO2 Absorption.* J. Phys. Chem. A 2009, 113, 6919−6923.

3.   Mejia-Trejo V. L.; Fregoso-Israel E.; Pfeiffer H. *Textural, Structural, and CO2 Chemisorption Effects Produced on the Lithium Orthosilicate by Its Doping with Sodium (Li4−xNaxSiO4).* Chem. Mater. 2008, 20, 7171−7176.

4.   Mosqueda H. A.; Vazquez C.; Bosch P.; Pfeiffer H. *Chemical Sorption of Carbon Dioxide (CO2) on Lithium Oxide (Li2O).* Chem. Mater. 2006, 18, 2307−2310.

5.   Shan S. Y.; Jia Q. M.; Jiang L. H.; Li Q. C.; Wang Y. M.; Peng J. H. *Novel Li4SiO4-Based Sorbents from Diatomite for High Temperature CO2 Capture.* Ceram. Int. 2013, 39, 5437−5441.

6.   Olivares-Marín M.; Castro-Díaz M.; Drage T. C.; Maroto-Valer M. M. *Use of Small-Amplitude Oscillatory Shear Rheometry to*

Study the Flow Properties of Pure and Potassium-Doped Li2ZrO3 Sorbents During the Sorption of CO2 at High Temperatures. Sep. Purif. Technol. 2010, 73, 415−420.

7. Pacciani R.; Torres J.; Solsona P.; Coe C.; Quinn R.; Hufton J.; Golden T.; Vega L. F. *Influence of the Concentration of CO2 and SO2 on the Absorption of CO2 by a Lithium Orthosilicate-Based Absorbent.* Environ. Sci. Technol. 2011, 45, 7083−7088.

8. Xiao Q.; Tang X.; Liu Y.; Zhong Y.; Zhu W. *Citrate Route to Prepare K-Doped Li2ZrO3 Sorbents with Excellent CO2 Capture Properties.* Chem. Eng. J. 2011, 174, 231−235.

9. Xiao Q.; Liu Y.; Zhong Y.; Zhu W. *A Citrate Sol-Gel Method to Synthesize Li2ZrO3 Nanocrystals with Improved CO2 Capture Properties.* J. Mater. Chem. 2011, 21, 3838−3842.

10. Rodríguez-Mosqueda R.; Pfeiffer H. *Thermokinetic Analysis of the CO2 Chemisorption on Li4SiO4 by Using Different Gas Flow Rates and Particle Sizes.* J. Phys. Chem. A 2010, 114, 4535−4541.

11. Ortiz-Landeros J.; Gomez-Yañez C.; Palacios-Romero L. M.; Lima E.; Pfeiffer H. *Structural and Thermochemical Chemisorption of CO2 on Li4+x(Si1−xAlx)O4 and Li4−x(Si1−xVx)O4 Solid Solutions.* J. Phys. Chem. A 2012, 116, 3163−3171.

12. Alcerreca-Corte I.; Fregoso-Israel E.; Pfeiffer H. *CO2 Absorption on Na2ZrO3: A Kinetic Analysis of the Chemisorption and Diffusion Processes.* J. Phys. Chem. C 2008, 112, 6520−6525.

13. Pfeiffer H.; Vazquez C.; Lara V. H.; Bosch P. *Thermal Behavior and CO2 Absorption of Li2-xNaxZrO3 Solid Solutions.* Chem. Mater. 2007, 19, 922−926.

14. Zhao T.; Ochoa-Fernández E.; Rønning M.; Chen D. *Preparation and High-Temperature CO2 Capture Properties of Nanocrystalline Na2ZrO3.* Chem. Mater. 2007, 19, 3294−3301.

15. Seggiani M.; Puccini M.; Vitolo S. *Alkali Promoted Lithium Orthosilicate for CO2 Capture at High Temperature and Low Concentration.* Int. J. Greenhouse Gas Control 2013, 17, 25−31.

16. Khokhani M.; Khomane R. B.; Kulkarni B. D. *Sodium-Doped Lithium Zirconate Nano-Squares: Synthesis, Characterization and Applications for CO2 Sequestration.* J. Sol-Gel Sci. Technol. 2012, 61, 316−320.

17. Veliz-Enriquez M. Y.; Gonzalez G.; Pfeiffer H. *Synthesis and CO2 Capture Evaluation of Li2−xKxZrO3 Solid Solutions and Crystal Structure of a New Lithium-Potassium Zirconate Phase*. J. Solid State Chem. 2007, 180, 2485−2492.

18. Martínez-dlCruz L.; Pfeiffer H. *Microstructural Thermal Evolution of the Na2CO3 Phase Produced During a Na2ZrO3−CO2 Chemisorption Process*. J. Phys. Chem. 2012, 116, 9675−9680.

19. Santillan-Reyes G. G.; Pfeiffer H. *Analysis of the CO2 Capture in Sodium Zirconate (Na2ZrO3). Effect of the Water Vapor Addition*. Int. J. Greenhouse Gas Control 2011, 5, 1624−1629.

20. Iwana A.; Stephenson H.; Ketchie C.; Lapkin A. *High Temperature Sequestration of CO2 Using Lithium Zirconates*. Chem. Eng. J. 2009, 146, 249−258.

21. Tabarés F. L. Editor, *Lithium: Technology, Performance and Safety*, Nova Publishers, (2013). Chapter 6, *Lithium Ceramics as an Alternative for the CO2 Capture. Analysis of Different Physicochemical Factors Controlling this Process*, pp 171-192.

22. Bish D.L., *Anion-Exchange in Takovite: Applications to Other Hydroxide Minerals*, Bone Miner., 1980, 103, 170-175.

23. Duan X.; Evans D. G., *Layered Double Hydroxides. Structure and Bonding*, Eds. Springer-Verlag: Berlin Heidelberg. Germany, 2006; vol. 119.

24. Wang M. Z.; Hu Q. D. L.; Li Y.; Li S.; Zhang X.; Xi M.; Yang X., *Intercalation of Ga3+-Salicylidene-Amino Acid Schiff Base Complexes into Layered Double Hydroxides: Synthesis, Characterization, Acid Resistant Property, in Vitro Release Kinetics and Antimicrobial Activity*, Appl. Clay Sci. 2013, 83&84, 182-190.

25. Catti M.; Ferraris G.; Hull S.; Pavese A. *Static Compression and H Disorder in Mg(OH)2 (Brucite) to 11 GPa: a Powder Neutron Diffraction Study*. Phys. Chem. Miner. 1995, 22, 200-206.

26. He J.; Wei M.; Li B.; Kang Y.; Evans D. G.; Duan X., *Preparation of Layered Double Hydroxides*, Structure and Bonding, 2005, 119, 89-119.

27. Gutmann N.; Müller B. *Insertion of the Dinuclear Dihydroxo-Bridged Cr(III) Aquo Complex into the Layered Double Hydroxides of Hydrotalcite-Type*. J. Solid State Chem. 1996, 122, 214-220.

28. Fogg A.M.; Williams G.R.; Chester R.; O'Hare D. *A Novel Family of Layered Double Hydroxides—[MgAl4(OH)12] (NO3)2 xH2O (M = Co, Ni, Cu, Zn)*. J. Mater. Chem., 2004, 14, 2369-2371.

29. de Roy A.; Forano C.; Besse J.P. *Layered Double Hydroxides: Synthesis and Post-Synthesis Modification*. Review, Chapter 1, 2002.

30. Choi S.; Drese J.H.; Jones C.W. *Adsorbent Materials for Carbon Dioxide Capture from Large Anthropogenic Point Sources*. ChemSusChem. 2009, 2, 796-854.

31. Ding Y.; Alpay, E. *Equilibria and Kinetics of CO2 Adsorption on Hydrotalcite Adsorbent*. Chem. Eng. Sci. 2000, 55, 3461-3474.

32. Reijers H.T.J.; Valster-Schiermeier S.E.A.; Cobden P.D.; van der Brink R.W. *Hydrotalcite as CO2 Sorbent for Sorption-Enhanced Steam Reforming of Methane*. Ind. Eng. Chem. Res. 2006, 45, 2522-2530.

33. Wang X.P.; Yu J.J.; Cheng J.; Hao Z.P.; Xu Z.P. *High Temperature Adsorption of Carbon Dioxide on Mixed Oxides Derived from Hydrotalcite-Like Compounds*. Environ. Sci. Technol. 2008, 42, 614-618.

34. Reynolds S.P.; Ebner A.D.; Ritter J.A. *Carbon Dioxide Capture from Flue Gas by Pressure Swing Adsorption at High Temperature Using a K-Promoted HTLC: Effects of Mass Transfer on the Process Performance*. Environ. Prog. 2006, 25, 334-342.

35. Reynolds, S.P.; Ebner, A.D.; Ritter, J.A. *Stripping PSA Cycles for CO2 Recovery from Flue Gas at High Temperature Using a Hydrotalcite Like Adsorbent*. Ind. Eng. Chem. Res. 2006, 45, 4278-4294.

36. Ebner, A.D.; Reynolds, S.P.; Ritter, J.A. *Nonequilibrium Kinetic Model that Describes the Reversible Adsorption and Desorption Behavior of CO2 in a K-Promoted Hydrotalcite-Like Compound*. Ind. Eng. Chem. Res. 2007, 46, 1737-1744.

37. Cavani F.; Trifirb F.; Vaccari A. *Hydrotalcite-Type Anionic Clays: Preparation, Properties and Applications*, Catal. Today, 1991, 11, 173–301.

38. Vaccari A. *Preparation and Catalytic Properties of Cationic and Anionic Clays*, Catal. Today, 1998, 41, 53–71.

39. Das N.N.; Konar J.; Mohanta M.K.; Srivastava S.C. *Adsorption of Cr(VI) and Se(IV) from their Aqueous Solutions onto Zr4+-Substituted ZnAl/MgAl-Layered Double Hydroxides: Effect of Zr4+ Substitution in the Layer*, J. Colloid Interf. Sci., 2004, 270, 1–8.

40. Goh K.H.; Lim T.T; Dong Z. *Application of Layered Double Hydroxides for Removal of Oxyanions: A Review*, Water Research, 2008, 42, 1343–1368.

41. Yong Z.; Mata, V.; Rodriguez, A.E. *Adsorption of Carbon Dioxide onto Hydrotalcite-Like Compounds (HTLCs) at High Temperatures*. Ind. Eng. Chem. Res. 2001, 40, 204-209.

42. Bellotto M.; Rebours B.; Clause O.; Lynch J.; Bazin D.; Elkaîm E. *Hydrotalcite Decomposition Mechanism: A Clue to the Structure and Reactivity of Spinel-Like Mixed Oxides*, J. Phys. Chem. 1996, 100, 8535-8542.

43. Ram Reddy M. K., Xu Z. P., Lu G. Q. (Max); Diniz da Costa J. C. *Layered Double Hydroxides for CO2 Capture: Structure Evolution And Regeneration*. Ind. Eng. Chem. Res. 2006, 45, 7504-7509.

44. Hufton J.; Mayorga S.; Gaffeney T.; Nataraj S.; Sircar S. *Sorption Enhanced Reaction Process (SERP)*, Proceedings of the 1997 U.S., DOE Hydrogen Program Review, 1997, 1, 179-194.

45. Ram Reddy M.K.; Xu Z.P.; Lu G.Q. (Max); Diniz da Costa J.C. *Influence of Water on High-Temperature CO2 Capture Using Layered Double Hydroxide Derivatives*. Ind. Eng. Chem. Res. 2008, 47, 2630-2635.

46. Martunus; Othman M.R; Fernando W.J.N. *Elevated Temperature Carbon Dioxide Capture Via Reinforced Metal Hydrotalcite*. Micropor. Mesopor. Mater. 2011, 138, 110–117.

47. Yong Z.; Rodrigues A.E. *Hydrotalcite-Like Compounds as Adsorbents for Carbon Dioxide*. Energy Convers. & Manage. 2002, 43, 1865-1876.

48. Newman S. P.; Jones W. *Supramolecular Organization and Materials Design*, Cambridge University Press, England, 2001.

49. Yamamoto T.; Kodama T.; Hasegawa N.; Tsuji M.; Tamura Y. *Synthesis of Hydrotalcite with High Layer Charge for CO2 Adsorbent*. Energy Convers Mgmt, 1995, 36, 637-640.

50. Wang Q.; Wu Z.; Tay H. H.; Chen L.; Liu Y.; Chang J.; Zhong Z.; Luo J.; Borgna A. *High Temperature Adsorption of CO2 on Mg-Al Hydrotalcite: Effect of the Charge Compensating Anions and the Synthesis pH*. Catal. Today, 2011, 164, 198-203

51. Wang Q.; Tay H.H.; Ng D.J.W.; Chen L.; Liu Y.; Chang J.; Zhong Z.; Luo J.; Borgna A. *The Effect of Trivalent Cations on the Performance of Mg-M-CO3 Layered Double Hydroxides for High-Temperature CO2 Capture*. ChemSusChem. 2010, 3, 965-973.

52. Hibino T.; Yamashita Y.; Kosuge K.; Tsunashima A. *Decarbonation Behavior of Mg-Al-CO3 Hydrotalcite-Like Compounds During Heat Treatment,*. Clays Clay Minerals. 1995, 43, 427 – 432.

53. Qian W.; Luo J.; Zhong Z.; Borgna A. *CO2 Capture by Solid Adsorbents and their Applications: Current Status and New Trends*. Energy Environ. Sci., 2011, 4, 42-55.

54. Hufton J. R.; Mayorga S.; Sircar S. *Sorption-Enhanced Reaction Process for Hydrogen Production*. AIChE J. 1999, 45, 248.

55. Oliveira E.L.G.; Grande C.A.; Rodrigues A.E.; *CO2 Sorption on Hydrotalcite and Alkali-Modified (K and Cs) Hydrotalcites at High Temperatures*. Sep. Purif. Technol. 2008, 62, 137-147.

56. Yang J. I.; Kim J. N. *Hydrotalcites for Adsorption of CO2 at High Temperature*. Korean J. Chem. Eng., 2006, 23, 77-80.

57. Ida J.I.; Lin S. *Mechanism of High-Temperature CO2 Sorption on Lithium Zirconate*. Environ. Sci. Technol., 2003, 37, 1999-2004.

58. Hufton J.; Mayorga S.; Nataraj S.; Sircar S.; Rao M. *Sorption-Enhanced Reaction Process (SERP)*, Proceedings of the 1998, USDOE Hydrogen Program Review, 1998, 2, 693-705.

59. Lee Jung M.; Min Yoon J.; Lee Ki B.; Jeon Sang G.; Na Jeong G.; Ryu Ho J. *Enhancement of CO2 Sorption Uptake on Hydrotalcite by Impregnation with K2CO3*. Langmuir, 2010, 26, 18788–18797.

60. Baker R. W. *Membrane Technology and Applications*, 2nd Editions, John Wiley and Sons (2004).

61. Murder M., *Basic Principles of Membrane Technology*, Kluwer Academic Publishers (1991).

62. Aresta M.; Dibenedetto A. *The Contribution of the Utilization Option to Reducing the CO2 Atmospheric Loading: Research Needed to Overcome Existing Barriers for a Full Exploitation of the Potential of The CO2 Use*, Cat. Today. 2004, 98, 455–462.

63. Anderson M.; Wang H.; Lin Y. S. *Inorganic Membranes for Carbon Dioxide and Nitrogen Separation*, Rev. Chem. Eng., 2012, 28, 101-121.

64. Yang H.; Xu Z.; Fan M.; Gupta R.; Slimane R. B.; Bland A. E.; Wright I. *Progress in Carbon Dioxide Separation and Capture: A Review*, J. Environ. Sci., 2008, 20, 14-27.

65. Niwa M.; Katada N.; Okumura K. *Introduction to Zeolite Science and Catalysis, Characterization and Design of Zeolite Catalysts*, Springer Series in Materials Science, Vol. 141 (2010).

66. Iwamoto Y.; Kawamoto H. *Science and Technology Trends: Quarterly Report*, 2009, 32, 42-59.

67. Algieri C.; Barbieri G.; Drioli E.; *Zeolite Membranes for Gas Separations, in Membrane Engineering for the Treatment of Gases*. Royal Society of Chemistry Vol. 2 (2011).

68. Fedosov D. A.; Smirnov A. V.; Knyazeva E. E.; Ivanova I. I. *Zeolite membranes: Synthesis, properties, and application*, Petroleum Chem., 2011, 51, 657-667.

69. Caro J.; Noack M. *Zeolite membranes – Recent developments and progress*. Micropor. Mesopor. Mater. 2008, 115, 215–233.

70. Jia M. D.; Peinemann K. V.; Behling R. D. *Ceramic Zeolite Composite Membranes. Preparation, Characterization and Gas Permeation*. J. Memb. Sci. 1993, 82, 15-26.

71. Cui Y.; Kita H.; Okamoto K. I. *Preparation and Gas Separation Performance of Zeolite T Membrane*. J. Mater. Chem. 2004, 14, 924.

72. Poshusta J.; Tuan V.; Pape E.; Noble R.; Falconer J. *Separation of Light Gas Mixtures Using SAPO-34 Membranes*. AIChE J. 2000, 46, 779-789.

73. Li S.; Falconer J.; Noble R. *SAPO-34 Membranes for CO2/CH4 Separation*. J. Memb. Sci., 2004, 241, 121–135.

74. Tomita T.; Nakayama K.; Sakai H. *Gas Separation Characteristics of DDR Type Zeolite Membrane*. Micropor. Mesopor. Mater. 2004, 68, 71.

75. Caro J.; Noack M. *Zeolite Membranes: Recent Developments and Progress*. Micropor. Mesopor. Mater. 2008, 115, 215-233.

76. Burggraaf A. J. *Fundamentals of Inorganic Membrane Science and Technology, Membrane Science and Technology*, Series, 4, Elsevier Science. Netherlands (1996).

77. Lara-Medina J. J.; Torres-Rodriguez M.; Gutierrez-Arzaluz M.; Mugica-Alvarez V. *Separation of CO2 and N2 with a Lithium-Modified Silicalite-1 Zeolite Membrane.* Inter. J. Greenhouse Gas Control, 2012, 10, 494-500.

78. Dyer A. *Introduction to Zeolite Science and Practice.* 3rd Revised Edition. J. Cejka, H. van Bekkum, A. Corma and F. Schiith (Editors) Elsevier B.V. (2007).

79. An W.; Swenson P.; Gupta A.; Wu L.; Kuznicki T. M.; Kuznicki S. M. *Improvement of H2/CO2 Selectivity of the Natural Clinoptilolite Membranes by Cation Exchange Modification.* J. Memb. Sci., 2013, 433, 25–31.

80. White J. C.; Dutta P. K.; Shqau K.; Verweij H. *Synthesis of Ultrathin Zeolite Y Membranes and their Application for Separation of Carbon Dioxide and Nitrogen Gases.* Langmuir 2010, 26, 12, 10287–10293.

81. Cho Y.K., Han K., Lee K.H. *Separation of CO2 by Modified -Al2O3 Membranes at High Temperature.* J. Membrane Sci. 1995, 104, 219-230.

82. Keizer K.; Uhlhorn R.J.R.; van Vuren R.J.; Burggraaf A.J. *Gas Separation Mechanisms in Microporous Modified -A12O3 Membranes.* J. Membrane Sci., 1988, 39, 285-300.

83. Uhlhorn R.J.R.; Keizer K., Burggraaf A.J. *Gas and Surface Diffusion in Modified 33-Alumina Systems.* J. Membrane Sci., 1989, 46, 225-241.

84. Kusakabe K., Ichiki K., Morooka S. *Separation of CO2 with BaTiO3 Membrane Prepared by the Sol—Gel Method.* J. Membrane Sci. 1944, 95, 171-177.

85. Nomura M., Sakanishi T., Nishi Y., Utsumi K., Nakamura R.. *Preparation of CO2 Permselective Li4SiO4 Membranes by Using Mesoporous Silica as a Silica Source.* Energy Procedia 2013, 37, 1004-1011.

86. Nomura M.; Nishi Y.; Sakanishi T.; Utsumi K.; Nakamura R. *Preparation of Thin Li4SiO4 Membranes by Using a CVD Method,* Energy Procedia 2013, 37, 1012-1019.

87. Kawamura H.; Yamaguchi T.; Nair B. N.; Nakagawa K.; Nakao S.I. *Dual-Ion Conducting Lithium Zirconate-Based Membranes for High Temperature CO2 Separation.* J. Chem. Eng. Jpn. 2005, 38, (5) 322-328.

88. Yamaguchi T.; Niitsume T.; Nair B. N.; Nakagawa K. *Lithium Silicate Based Membranes for High Temperature CO2 Separation.* J. of Membrane Sci., 2007, 294, 16-21.

89. Nair B.N.; Burwood R.P.; Goh V.J.; Nakagawa K.; Yamaguchi T. *Lithium Based Ceramic Materials and Membranes for High Temperature CO2 Separation.* Progress in Materials Sci. 2009, 54, 511–541.

90. Skinner S.J.; Kilner J.A. *Oxygen Ion Conductors.* Materials Today, 2003, 6, 30-37.

91. Anderson M.; Lin Y.S. *Carbonate–Ceramic Dual-Phase Membrane for Carbon Dioxide Separation.* J. Membrane Sci. 2010, 357, 22.

92. Wade J. L.; Lee C.; West A. C.; Lackner K. S. *Composite Electrolyte Membranes for High Temperature CO2 Separation.* J. Membrane Sci. 2011, 369, 20.

93. Zhang L.; Xu N.; Li X.; Wang S.; Huang K.; Harris W. H.; Wilson K.; Chiu S. *High CO2 Permeation Flux Enabled by Highly Interconnected Three-Dimensional Ionic Channels in Selective CO2 Separation Membranes.* Energy Environ. Sci. 2012, 5, 8310.

94. Rui Z.; Anderson M.; Li Y.; Lin Y.S. *Ionic Conducting Ceramic and Carbonate Dual Phase Membranes for Carbon Dioxide Separation,* J. Membrane Sci. 2012, 417-418, 174.

95. Lu B.; Lin Y.S. *Synthesis and Characterization of Thin Ceramic-Carbonate Dual-Phase Membranes for Carbon Dioxide Separation.* J. Membrane Sci. 2013, 444, 402–411.

96. Dong X.; Ortiz-Landeros J.; Lin Y. S. *An Assymetric Thin Tubular Dual Phase Membrane.* Chem. Commun. 2013, 49, 9654.

97. Ortiz-Landeros J.; Norton T.T.; Lin Y.S. *Effects of Support Pore Structure on Carbon Dioxide Permeation of Ceramic-Carbonate Dual-Phase Membranes.* Chem. Eng. Sci. 2013, 104, 891-898.

98. Rui Z.; Anderson M.; Lin Y.S.; Li Y.; *Modeling and Analysis of Carbon Dioxide Permeation through Ceramic-Carbonate Dual-Phase Membranes.* J. Membrane Sci. 2009, 345, 110.

99. Yu K.M.K.; Curcic I.; Gabriel J.; Tsang S.C.E. *Recent Advances in CO2 Capture and Utilization,* ChemSusChem. 2008, 1, 893–899.

100. Wender I. *Reactions of Synthesis Gas,* Fuel Process. Technol. 1996, 48, 3, 189-297.

101. Rostrup-Nielsen J. R. *Syngas in Perspective*. Catal. Today 2002, 71, 3-4, 243-247.

102. Dry M.E. *The Fischer–Tropsch Process: 1950–2000*. Catal. Today 2002, 71, 227-241.

103. Yu K.M.K.; Curcic I.; Gabriel J.; Tsang S.C.E. *Recent Advances in CO2 Capture and Utilization*. ChemSusChem. 2008, 1, 893 – 899.

104. Gnanapragasam N.; Reddy B.; Rosen M. *Reducing CO2 Emissions for an IGCC Power Generation System: Effect of Variations in Gasifier and System Operating Conditions*. Energ. Convers. Manage. 2009, 50, 1915-1923.

105. Yang Z.Z.; He L.N.; Gao J.; Liu A.H.; Yu B. *Carbon Dioxide Utilization with C–N Bond Formation: Carbon Dioxide Capture and Subsequent Conversion*. Energy Environ. Sci. 2012, 5, 6602-6639.

106. Chattopadhyay S.; Sen R. *Fuel Properties, Engine Performance and Environmental Benefits of Biodiesel Produced by a Green Process*. Appl. Energ. 2013, 105, 319–326.

107. Choi J.C.; He L.N.; Yasuda H.; Sakakura T. *Selective and High Yield Synthesis of Dimethyl Carbonate Directly from Carbon Dioxide and Methanol*. Green Chem. 2002, 4, 230–234.

108. Quispe C.A.; Coronado J.R.C; CarvalhoJr J.A. *Glycerol: Production, Consumption, Prices, Characterization and New Trends in Combustion*.Renew. Sust. Energ. Rev. 2013, 27, 475–493.

109. Talebian-Kiakalaieh A.; Saidina Amin N. A; Mazaheri H. *A Review on Novel Processes of Biodiesel Production from Waste Cooking Oil*. Appl. Energ. 2013, 104, 683–710.

110. Kuwahara Y.; Tsuji K.; Ohmichi T.; Kamegawa T.; Moria K.; Yamashita H. *Transesterifications Using a Hydrocalumite Synthesized from Waste Slag: An Economical and Ecological Route for Biofuel Production*. Catal. Sci. Tech. 2012, 2, 1842–1851.

111. Atadashi I.M.; Aroua M.K.; Abdul Aziz A. *Biodiesel Separation and Purification: A Review*. Renew. Energ. 2011, 36, 437-443.

112. Lam M.K.; Lee K.T. *Mixed Methanol–Ethanol Technology to Produce Greener Biodiesel from Waste Cooking Oil: A*

Breakthrough for SO42⁻/SnO2–SiO2 catalyst. Fuel Process. Technol. 2011, 92, 1639–1645.

113. Lam M.; Lee K.T.; Rahman Mohamed A. *Homogeneous, Heterogeneous and Enzymatic Catalysis for Transesterification of High Free Fatty Acid Oil (Waste Cooking Oil) to Biodiesel: A Review.* Biotechnol. Adv. 2010, 28, 500–518.

114. Georgogianni K.G.; Katsoulidis A.K.; Pomonis P.J.; Manos G.; Kontominas M.G. *Transesterification of Rapeseed Oil for the Production of Biodiesel Using Homogeneous and Heterogeneous Catalysis.* Fuel Process. Technol. 2009, 90, 1016–1022.

115. Zhang Y.; Dube M.A.; McLean D.D.; Kates M. *Biodiesel Production from Waste Cooking Oil: 1. Process Design and Technological Assessment.* Bioresource Technol. 2003, 89, 1-16.

116. Long Y.D.; Fang Z.; Su T.C.; Yang Q. *Co-production of Biodiesel and Hydrogen from Rapeseed and Jatropha Oils with Sodium Silicate and Ni Catalysts.* Appl. Energ. 2013, http://dx.doi.org/10.1016/j.apenergy.2012.12.076.

117. Centi G.; Quadrelli E.A.; Perathoner S. *Catalysis for CO2 Conversion: A Key Technology for Rapid Introduction of Renewable Energy in the Value Chain of Chemical Industries.* Energy Environ. Sci. 2013, 6, 1711-1731.

118. Centi G.; Perathoner S. *Opportunities and Prospects in the Chemical Recycling of Carbon Dioxide to Fuels.* Catal. Today 2009, 148, 191-205.

119. Hoffmann F.M.; Yang Y.; Paul J.; White M.G; Hrbek J. *Hydrogenation of Carbon Dioxide by Water: Alkali-Promoted Synthesis of Formate.* J. Phys. Chem. Lett. 2010, 1, 2130–2134.

120. H. Yin, X. Mao, D. Tang, W. Xiao, L. Xing, H. Zhu, D. Wang, D.R. Sadoway. *Capture and Electrochemical Conversion of CO2 to Value-Added Carbon and Oxygen by Molten Salt Electrolysis.* Energy Environ. Sci., 2013, 6, 1538-1545.

121. Hu B.; Guild C.; Suib S.L. *Thermal, Electrochemical, and Photochemical Conversion of CO2 to Fuels and Value-Added Products.* Journal of CO2 Utilization. 2013, 1, 18–27.

122. Kumar B.; Smieja J.M.; Kubiak C.P. *Photo-reduction of CO2 on p-type Silicon Using Re(Bipy-But)(CO)3Cl: Photovoltages Exceeding 600 mv for the Selective Reduction of CO2 to CO.* J. Phys. Chem. C 2010, 114, 14220–14223.

123. Bara J. *Review: The Chemistry of Amine Manufacture. What Chemicals Will We Need to Capture CO2?* Greenhouse Gas Sci. Technol. 2012, 2, 162–171.

124. Helwani Z.; Othman M.R.; Aziz N.; Fernando W.J.N.; Kim J. *Technology for Production of Biodiesel Focusing on Green Catalytic Techniques: A Review.* Fuel Process. Technol. 2009, 90, 1502–1515.

125. Endalew K.; Kiros Y., Zanzi R. *Inorganic Heterogeneous Catalysts for Biodiesel Production from Vegetable Oils.* Biomass Bioenerg. 2011, 35, 3787-3809.

126. Luque R.; Lovett J.C.; Datta B.; Clancy J.; Campeloa J.M.; Romero A.A. *Biodiesel as Feasible Petrol Fuel Replacement: A Multidisciplinary Overview.* Energy Environ. Sci. 2010, 3, 1706–1721.

127. Ramachandran K.; Suganya T.; Gandhi N.N.; Renganathan S. *Recent Developments for Biodiesel Production by Ultrasonic Assist Transesterification Using Different Heterogeneous Catalyst: A Review.* Renew. Sust. Energ. Rev. 2013, 22, 410–418.

128. Vyas A.P.; Verma J.L.; Subrahmanyam N. *A Review on FAME Production Processes.* Fuel 2010, 8, 1–9.

129. Di Serio M., Ledda M., Cozzolino M., Minutillo G., Tesser R., Santacesaria E. *Transesterification of Soybean Oil to Biodiesel by Using Heterogeneous Basic Catalysts.* Ind. Eng. Chem. Res. 2006, 45, 3009-3014.

130. Veljkovic V. B.; Stamenkovic O. S.; Todorovic Z. B.; Lazic M. L.; Skala D. U. *Kinetics of Sunflower Oil Methanolysis Catalyzed by Calcium Oxide.* Fuel 2009, 88, 554–1562.

131. Singh Chouhan A.P.; Sarma A.K. *Modern Heterogeneous Catalysts for Biodiesel Production: A Comprehensive Review.* Renew. Sust. Energ. Rev. 2011, 15, 4378– 4399.

132. Lee J.S.; Saka S. *Biodiesel Production by Heterogeneous Catalysts and Super-Critical Technologies: Review.* Bioresource Technol. 2010, 101, 7191–7200.

133. Salamatinia B.; Abdullah A.Z.; Bhatia S. *Quality Evaluation of Biodiesel Produced through Ultrasound-Assisted Heterogeneous Catalytic System.* Fuel Process. Technol. 2012, 97, 1-8.

134. Berrios M.; Martín M.A.; Chica A.F.; Martín A. *Purification of Biodiesel from Used Cooking Oils*. Appl. Energ. 2011, 88, 3625–3631.

135. Zabeti M.; Daud W.M.A.W.; Aroua M.K. *Activity of Solid Catalysts for Biodiesel Production: A Review*. Fuel Process. Technol. 2009, 90, 770–777.

136. Sharma Y. C.; Singh B.; Korstad J. *Latest Developments on Application of Heterogenous Basic Catalysts for an Efficient and Eco Friendly Synthesis of Biodiesel: A Review*. Fuel 2011, 90, 1309–1324.

137. Wen Z.; Yu X.; Tu S.T.; Yan J.; Dahlquist E. *Synthesis of Biodiesel from Vegetable Oil with Methanol Catalyzed by Li-Doped Magnesium Oxide Catalysts*. Appl. Energ. 2010, 87, 743–748.

138. Dossin T.F.; Reyniers M.F.; Berger R.J.; Marin G.B. *Simulation of Heterogeneously MgO-Catalyzed Transesterification for Fine-Chemical and Biodiesel Industrial Production*. Appl. Catal. B-Environ. 2006, 67, 136–148.

139. Liu X.; He H.; Wang Y.; Zhu S.; Piao X. *Transesterification of Soybean Oil to Biodiesel Using CaO as a Solid Base Catalyst*. Fuel 2008, 87, 216–21.

140. Liu X.; He H.; Wang Y.; Zhu S. *Transesterification of Soybean Oil to Biodiesel Using SrO as a Solid Base Catalyst*. Catal. Commun. 2007, 8, 1107–1111.

141. Soares Días A.P.; Bernardo J.; Felizardo P.; Neiva Correia M.J. *Biodiesel Production by Soybean Oil Methanolysis over SrO/ MgO Catalysts. The Relevance of the Catalyst Granulometry*. Fuel Process. Technol. 2012, 102, 146–155.

142. Granados M.L.; Poves M.D.; Alonso D.; Mariscal R.; Galisteo F.C.; Moreno-Tost R. *Biodiesel from Sunflower Oil by Using Activated Calcium Oxide*. Appl. Catal. B-Environ. 2007, 73, 317–26.

143. Watkins R.S.; Lee A.F.; Wilson K. *Li–CaO Catalysed Tri-Glyceride Transesterification for Biodiesel Applications*. Green Chem. 2004, 6, 335–340.

144. Wang J.X.; Chen K.T.; Huang J.S.; Chen C.C. *Application of Li2SiO3 as a Heterogeneous Catalyst in the Production of Biodiesel from Soybean Oil*. Chinese Chem. Lett. 2011, 22, 1363–1366.

145. Wang J.X.; Chen K.T.; Wu J.S.; Wang P.H.; Huang S.T.; Chen C.C. *Production of Biodiesel through Transesterification of Soybean Oil Using Lithium Orthosilicate Solid Catalyst.* Fuel Process. Technol. 2012, 104, 167–173.

146. Chen K.T.; Wang J.X.; Dai Y.M.; Wang P.H.; Liou C.Y.; Nien C.W.; Wu J.S.; Chen C.C. *Rice Husk Ash as a Catalyst Precursor for Biodiesel Production.* J. Taiwan Inst. Chem. Eng. 2013, 44, 622-629.

147. Guo P.; Zheng C.; Zheng M.; Huang F.; Li W.; Huang Q. *Solid Base Catalysts for Production of Fatty Acid Methyl Esters.* Renew. Energ. 2013, 53, 377-383.

148. Singh A.K.; Fernando S.D. *Transesterification of Soyabean Oil Using Heterogenous Catalysts.* Energ. Fuels 2008, 22, 2067-2069.

149. Omar W.N.N.W.; Amin N.A.S. *Biodiesel Production from Waste Cooking Oil over Alkaline Modified Zirconia Catalyst.* Fuel Process. Technol. 2011, 92, 2397-2405.

150. Molaei Dehkordi A.; Ghasemi M. *Transesterification of Waste Cooking Oil to Biodiesel Using Ca and Zr Mixed Oxides as Heterogeneous Base Catalysts.* Fuel Process. Technol. 2012, 97, 45–51.

151. Shibasaki-Kitakawa N., Honda H., Kuribayashi H., Toda T., Fukumura T., Yonemoto T. *Biodiesel Production Using Anionic Ion-Exchange Resin as Heterogeneous Catalyst.* Bioresource Technol. 2007, 98, 416–421.

152. Shumaker J. L.; Crofcheck C.; Tackett S.A.; Santillan-Jimenez E.; Morgan T.; Ji Y.; Mark Crocker; Toops T.J. *Biodiesel Synthesis Using Calcined Layered Double Hydroxide Catalysts.* Appl. Catal. B-Environ. 2008, 82, 120–130.

153. Corma A., Hamid S.B.A., Iborra S., Velty A. *Lewis and Bronsted Basic Active Sites on Solid Catalysts and their Role in the Synthesis of Monoglycerides.* J. Catal. 2005, 234, 340-347.

154. Sankaranarayanan S.; Churchil Antonyraj A.; Kannan S. *Transesterification of Edible, Non-Edible and used Cooking Oils for Biodiesel Production Using Calcined Layered Double Hydroxides as Reusable Base Catalysts.* Bioresource Technol. 2012, 109, 57–62.

155. Navajas A.; Campo I.; Arzamendi G.; Hernandez W.Y.; Bobadilla L.F.; Centeno M.A.; Odriozola J.A.; Gandia L.M. *Synthesis of Biodiesel from the Methanolysis of Sunflower Oil Using PURAL® Mg–Al Hydrotalcites as Catalyst Precursors.* Appl. Catal. B-Environ 2010, 100, 299–309.

156. Gomes J.F.P.; Puna J.F.B.; Gonçalves L.M.; Bordado J.C.M. *Study on the use of MgAl Hydrotalcites as Solid Heterogeneous Catalysts for Biodiesel Production.* Energy 2011, 36, 6770-6778.

157. Sánchez-Cantú M.; Pérez-Díaz L.M.; Tepale-Ochoa N.; González-Coronel V.J.; Ramos-Cassellis M.E.; Machorro-Aguirre D.; Valente J.S. *Green Synthesis of Hydrocalumite-Type Compounds and their Evaluation in the Transesterification of Castor Bean Oil and Methanol.* Fuel 2013, 111, 23-31.

158. Sun H.; Ding Y.; Duan J.; Zhang Q.; Wang Z.; Lou H.; Zheng X. *Transesterification of Sunflower Oil to Biodiesel on ZrO2 Supported La2O3 Catalyst.* Bioresource Technol. 2010, 101, 953–958.

159. Kim H.J.; Kang B.S.; Kim M.J.; Park Y.M.; Kim D.K.; Lee J.S.; Lee K.Y. *Transesterification of Vegetable Oil to Biodiesel Using Heterogeneous Base Catalyst.* Catal. Today 2004, 93–95, 315–320.

160. Ebiura T.; Echizen T.; Ishikawa A.; Murai K.; Baba T. *Selective Transesterification of Triolein with Methanol to Methyl Oleate and Glycerol Using Alumina Loaded with Alkali Metal Salt as a Solid-Base Catalyst.* Appl. Catal. A-Gen. 2005, 283, 111–116.

161. Xie W.; Li H. *Alumina-Supported Potassium Iodide as a Heterogeneous Catalyst for Biodiesel Production from Soybean Oil.* J. Mol. Catal. A-Chem. 2006, 255, 1–9.

162. Lukic I.; Krstic J.; Jovanovic D.; Skala D. *Alumina/Silica Supported K2CO3 as a Catalyst for Biodiesel Synthesis from Sunflower Oil.* Bioresource Technol. 2009, 100, 4690–4696.

163. Evangelista J.P.C.; Chellappa T.; Coriolano A.C.F.; Fernandes Jr. V.J.; Souza L.D.; Araujo A.S.; *Synthesis of Alumina Impregnated with Potassium Iodide Catalyst for Biodiesel Production from Rice Bran Oil.* Fuel Process. Technol. 2012, 104, 90–95.

164. Arzamendi G.; Campo I.; Arguiñarena E.; Sánchez M.; Montes M.; Gandía L.M. *Synthesis of Biodiesel with Heterogeneous*

NaOH/Alumina Catalysts: Comparison with Homogeneous NaOH. Chem. Eng. J. 2007, 134, 123–130.

165. Baroutian S.; Aroua M.K.; Raman A.A.; Sulaiman N.M.N. Methanol Recovery During Transesterification of Palm Oil in a TiO2/Al2O3 Membrane Reactor: Experimental Study and Neutral Network Modeling. Sep. Purif. Technol. 2010, 76, 58–63.

166. Wu H.; Zhang J.; Wei Q.; Zheng J.; Zhang J. Transesterification of Soybean Oil to Biodiesel Using Zeolite Supported CaO as Strong Base Catalysts. Fuel Process. Technol. 2013, 109, 13-18.

167. Babajide O.; Musyoka N.; Petrik L.; Ameer F. Novel Zeolite Na-X Synthesized from Fly Ash as a Heterogeneous Catalyst in Biodiesel Production.Catal. Today 2012, 190, 54-60.

168. Alves C.T.; Oliveira A.; Carneiro S.A.V; Silva A.G.; Andrade H.M.C.; Vieira de Melo S.A.B.; Torres E.A. Transesterification of Waste Frying Oil Using a Zinc Aluminate Catalyst. Fuel Process. Technol. 2013, 106, 102–107.

169. Borges M.E.; Díaz L. Recent Developments on Heterogeneous Catalysts for Biodiesel Production by Oil Esterification and Transesterification Reactions: A Review. Renew. Sust. Energ. Rev. 2012, 6, 2839–2849.

170. Di Serio M.; Tesser R.; Pengmei L.; Santacesaria E. Heterogeneous Catalysts for Biodiesel Production. Energ. Fuels 2008, 22, 207–17.

171. Wen Z., Yu X., Tu S.T., Yan J., Dahlquist E. Biodiesel Production from Waste Cooking Oil Catalyzed by TiO2–MgO Mixed Oxides. Bioresource Technol. 2010, 101, 9570–9576.

172. Guo F.; Peng Z.G.; Dai J.Y.; Xiu Z.L. Calcined Sodium Silicate as Solid Base Catalyst for Biodiesel Production. Fuel Process. Technol. 2010, 991, 322–328.

173. Singh A.K.; Fernando S.D.; Hernandez R. Base Catalyzed Fast Transesterification of Soybean Oil Using Ultrasonication. Energ. Fuel 2007, 21, 1161–1164.

174. Kalva A.; Sivasankar T.; Moholkar V.S. Physical Mechanism of Ultrasound-assisted Synthesis of Biodiesel. Ind. Eng. Chem. Res. 2008, 48, 534–544.

175. Nishimura CSMV; Maeda Y.R. Conversion of Vegetable Oil to Biodiesel Using Ultrasonic Irradiation. Chem. Lett. 2003, 32, 716–717.

176. Salamatinia B., Mootabadi H., Bhatia S., Abdullah A.Z. *Optimization of Ultrasonic-Assisted Heterogeneous Biodiesel Production from Palm Oil: A Response Surface Methodology Approach.* Fuel Process. Technol. 2010, 91, 441–448.

177. Mootabadi H.; Salamatinia B.; Bhatia S.; Abdullah A.Z. *Ultrasonic-assisted Biodiesel Production Process from Palm Oil Using Alkaline Earth Metal Oxides as the Heterogeneous Catalysts.* Fuel 2010, 89, 1818–1825.

178. Deng X.; Fang Z.; Liu Y.; Yu C. *Production of Biodiesel from Jatropha Oil Catalyzed by Nanosized Solid Basic Catalyst.* Energy 2011, 36, 777–784.

179. Deng X.; Fang Z.; Hu Y.; Zeng H.; Liao K.; Yu C.L. *Preparation of Biodiesel on Nano Ca–Mg–Al Solid Base Catalyst under Ultrasonic Radiation in Microaqueous Media.* Petrochemical Technology 2009, 38, 1071–1075.

Chapter 3

Physical Properties of G-class Cement for Geothermal Well Cementing in South Korea

Jongmuk Won[a], Dongseop Lee[b], Kyunguk Na[b],
In-Mo Lee[b], and Hangseok Choi[b]

[a]Georgia Institute of Technology, School of Civil & Environmental Engineering, USA

[b]Korea University, School of Civil, Environmental, & Architectural Engineering, South Korea

ABSTRACT

The cement material adopted for a new geothermal well project in South Korea is specialized as the G-class cement, which is commonly used in the oil-well industry, and regulated by the API (American Petroleum Institute). In order to maintain the optimal generating performance of geothermal wells, physical properties of the cementing material should be satisfactory. In this paper, the significant material properties (i.e., groutability, uniaxial compressive strength, thermal conductivity, bleeding potential, phenolphthalein indication) of the G-class cement were experimentally examined, with consideration of various water–cement (w/c) ratios as mix proportion. Important findings through the experiments are as follows; (1) Groutability of the G-class cement increases with the addition of a small amount of retarder. (2) There would be a structural problem when the w/c ratio is kept extremely high in order to obtain acceptable groutability. (3) Thermal conductivity of the G-class cement is small enough to prevent heat loss during circulating up hot steam or water from the deep underground to the ground surface. (4) The G-class cement used for geothermal-well cementing causes no bleeding problem. (5) The phenolphthalein indicator is applicable to distinguishing the G-class cement from the drilling mud.

INTRODUCTION

As a promising alternative to conventional fossil-based energy sources, geothermal energy is playing a growing role in supplying heat energy and electricity for human needs. Geothermal energy is defined as natural heat latent in the Earth's crust that is trapped close enough to the surface to be economically extracted. Compared to the direct use of geothermal energy, which is usually used for heating and cooling for residential or commercial buildings, electricity generation from a geothermal well is the most promising utilization of high-temperature (higher than 150 °C) geothermal resources, and mainly takes place in conventional steam turbines and binary plants, depending on the characteristics of the geothermal resources. Geothermal wells drilled for producing hot water or steam, and consequently for generating electricity are similar, in many respects, to oil wells [10].

In the future, the geothermal power plant will be more frequently constructed in combination with the EGS (Enhanced Geothermal System) technique as shown in Fig. 1, and the binary cycle power plant, replacing the conventional geothermal power plant in which electricity is directly generated from the hot water or steam yielded in favorable geothermal reservoirs. In particular, there is no high quality natural geothermal resource such as volcanic activity in the Korean Peninsula, but the EGS technique can be alternatively considered for operating geothermal power plants. The EGS concept is based on fracturing of a large volume of Hot Dry Rock (HDR) by injecting water, and the subsequent recirculation of water between a surface heat exchanger and the newly created artificial reservoir.

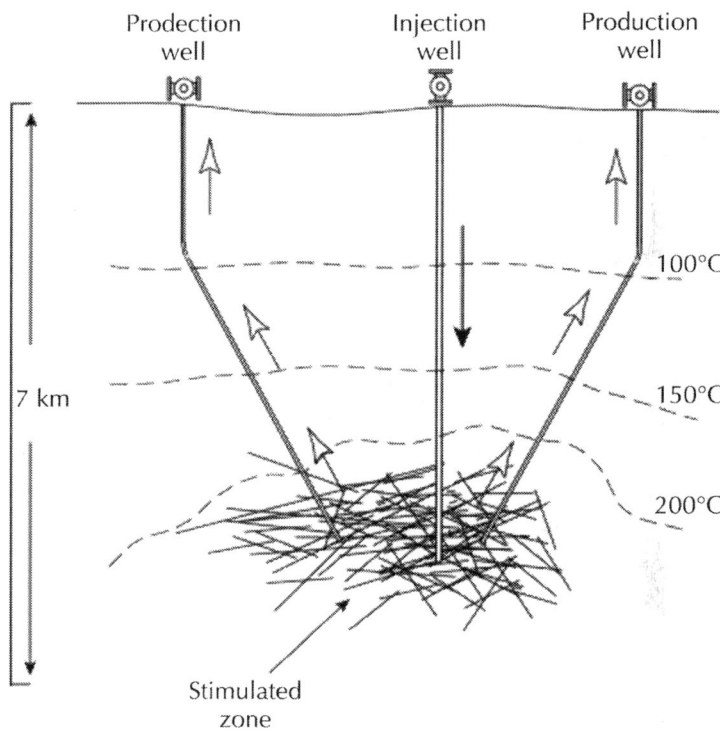

Figure 1: Schematics of EGS system [16].

In order to construct a geothermal power plant successfully, a complete cementing in geothermal wells is of particular importance for the following reasons: the hardened-cement under the ground must have sufficient strength to support the steel casing, and to avoid excessive temperature-induced deformation. The cement slurry should pre-flash and displace the water-based drilling fluids [3]. In addition, the complete cementing is to protect the steel casing from corrosive fluids. As a result, the cement material for completing geothermal wells mechanically supports steel casings, as well as protects from initial corrosion or erosion by geothermal fluids that are up to 320 °C maximum

[14]. The cement material used in geothermal wells is specialized as the G-class cement that is commonly adopted in the oil-well industry, which is regulated by the API (American Petroleum Institute). The G-class cement was developed to ensure sufficient durability at both high-temperature and high-pressure in deep geothermal reservoirs.

In this paper, the groutability, uniaxial compressive strength, thermal conductivity, and bleeding potential (i.e., free fluid content) of the G-class cement that is adopted in the first geothermal well project in South Korea were experimentally evaluated; these properties are crucial to satisfying the required function of geothermal well cementing. Each experiment was performed under various water–cement (w/c) ratios.

GROUTABILITY OF CEMENT SLURRY

Experimental Equipment and Procedure

Groutability of the cement slurry is a critical factor in the geothermal well design to deliver cement slurry successfully into the underground, without forming any faulty gaps between the casing and the ground formation. If the groutability of cement is insufficient, the cement slurry would be hardened early during the injection, hampering the continuous supply of cement. On the other hand, an excessively low viscosity of the cement slurry that attempts to enhance groutability leads to a loss of a large amount of cement slurry into cracks developed around the geothermal well, or problems of structural stability may occur. So, it is important to ensure the optimal groutability of cementing, considering geologic characteristics around a geothermal well.

In order to evaluate the groutability of G-class cement in this paper, the V-funnel test and the Slump flow test [6] were adopted that are recommended by the EFNARC (European Federation for Specialist Construction Chemicals and Concrete System) and JSCE (Japan Society of Civil Engineers). A typical mixture design for a cement slurry was prepared according to the mixture design of 40SF type (Table 1), proposed by Philippacopoulos and Berndt (2000). [9].

Table 1: Typical cement slurry mixture design of 40SF type (by mass)

Mix type	Cement	Silica flour	Water	Bentonite	Dispersant
40SF	1	0.4	0.55	0.034	0.012

The groutability of G-class cement was experimentally evaluated with consideration of various water–cement (w/c) ratios (i.e., 0.55, 0.6, 0.7, 0.8) of the cement slurry and the addition of a retarder to the 40SF type mixture design along with various retarder–cement (r/c) ratios from 0.005 to 0.025.

In the V-funnel test, the cement slurry should be carefully fed into a V-funnel with a volume 12 L (Fig. 2), held in the container with no leakage for 10 s, and allowed to pour down by opening the exit. A small amount of water was spread on the inner surface of the V-funnel before pouring the cement slurry to minimize frictional resistance between the slurry and the acrylic V-funnel. One should record the time required for all of the cement slurry removed from the V-funnel.

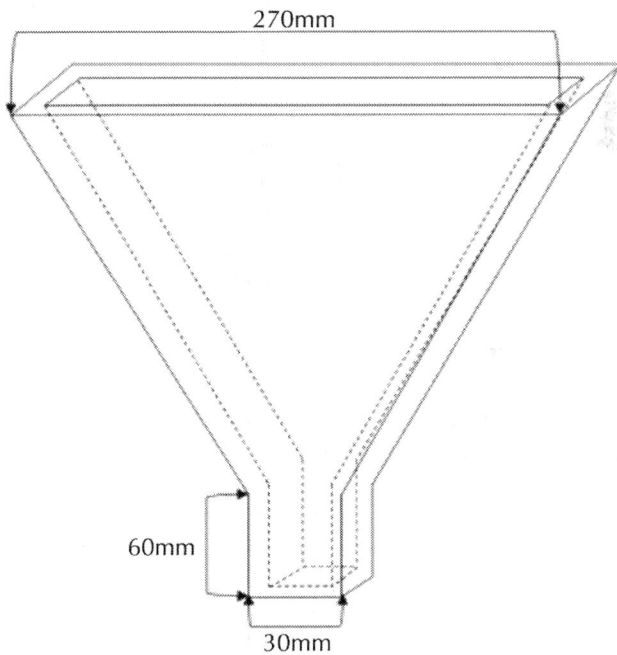

Figure 2: Schematics of V-funnel test.

A series of the V-funnel tests was carried out for six different holding times in the V-funnel (i.e., 0, 10, 20, 30, 40, 50 min), after placing the cement slurry, to assess an increase in viscosity with curing time. The groutability estimated from the V-funnel test is quantified as the value of R_m (relative funnel speed of mortar), as defined in Eq. (1) [4].

$$R_m = \frac{10}{t}$$

(1)

where   t   (sec)   is   the   time   to

Whereas the slump flow test measures the groutability (or flowability) of cement slurry in the transverse direction, unlike the V-funnel test that estimates the groutability in the vertical or gravitational direction. In the process of the slump flow test, a 60-mm-high tapered-brass ring with diameters of 100 mm at the bottom and of 70 mm at the top is placed on an acrylic plate, and is filled with the cement slurry as shown in Fig. 3. After lifting up the brass ring, the maximum and minimum diameters of the cement slurry spreading over the acrylic place should be recorded to determine the groutability. In this paper, the slump flow tests were carried out for six different holding times in the brass ring (i.e., 0, 10, 20, 30, 40, 50 min), after placing the cement slurry, to assess an increase in viscosity with curing time. The groutability of cement slurry through the slump flow test is quantified as the value of c (relative flow area of mortar), that is defined as Eq. (2)[4].

$$\Gamma_c = \frac{d_1 d_2 - d_0^2}{d_0^2}$$

(2)

where d_1 (cm) and d_2 (cm) are the diameter of spreading cement slurry in the longitudinal and transverse direction, respectively. d_0 (cm) is the bottom diameter of the brass ring.

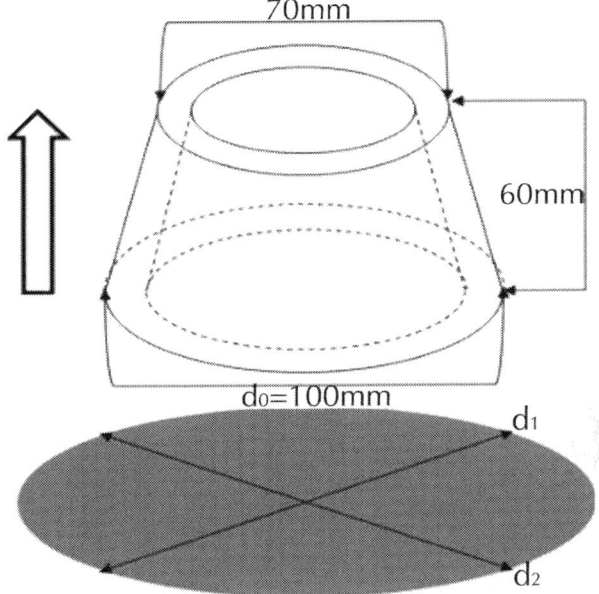

Figure 3: Schematics of slump flow test.

Comparison of Groutability

The groutability of G-class cement slurry estimated by both the V-funnel test and slump flow test is summarized in Fig. 4 for the water–cement (w/c) ratio of 0.7 and in Fig. 5 for the water–cement (w/c) ratio of 0.8. The values of R_m and Γc tend to decrease with increasing curing time in the containers, which indicates the viscosity of G-class cement slurry rapidly increases with time. In addition, the experimental results for the w/c ratios of 0.55, 0.6, 0.7 and 0.8 are compared in the R_m-Γc chart as shown in Fig. 6. The slump flow test result (Γc) is represented on the x-axis, and the V-funnel test result (R_m) is represented on the y-axis. The values R_m and Γc are in proportion to each other, and increase with increasing w/c ratio. The largest values of R_m and Γc represent the holding (or curing) time of 0 min, and the smallest values represent the holding time of 50 min. However, the groutability of G-class cement slurry without adding a retarder is overall very low. Moreover, the w/c ratios of 0.55 and 0.6 show the R_m and Γc values almost zero as noted in Fig. 6.

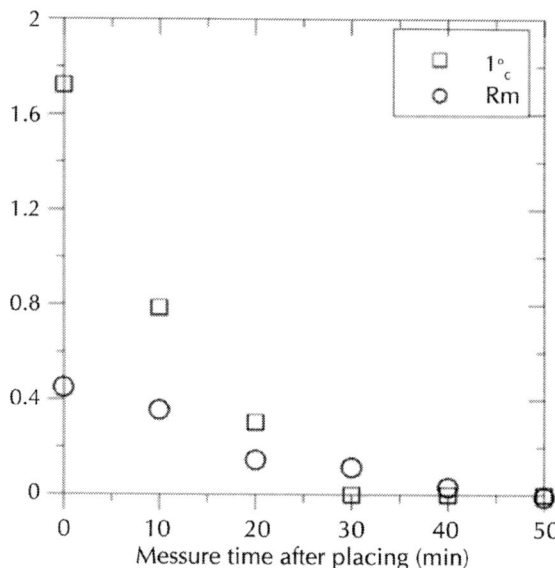

Figure 4: Groutability test results for w/c = 0.7.

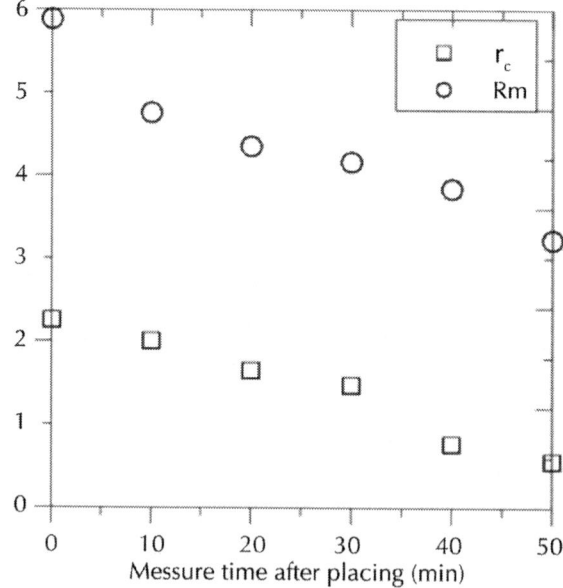

Figure 5: Groutability test results for w/c = 0.8.

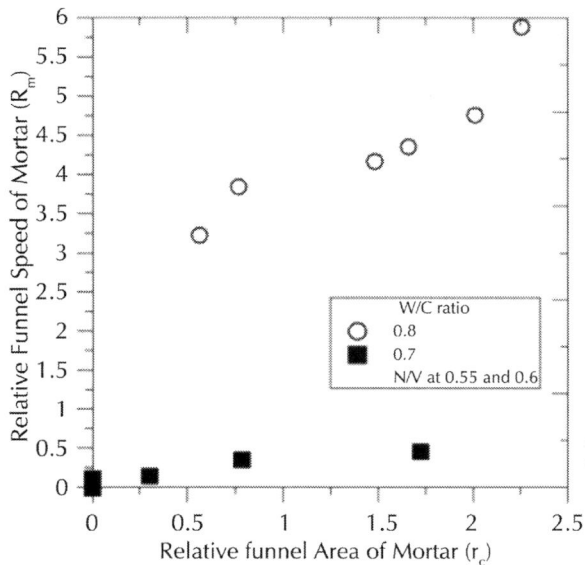

Figure 6: Groutability of G-class cement slurry for various w/c ratios without retarder.

Based on the experimental results, the w/c ratio of 0.8 can be selected as an optimum mixture design for the G-class cement that provides enough groutability suitable for a geothermal well cementing. However, this w/c ratio is rather high compared to the mixture design of 40SF type proposed by Philippacopoulos and Berndt (2000) [9]. Such a high w/c ratio may reduce the cementing strength, and thus cause structural integrity problems.

In order to maintain the acceptable groutability of G-class cement slurry with an adequate w/c ratio, a retarder (Halliburton Energy Services) was added to the G-class cement slurry. The w/c ratio was fixed as 0.55 conforming to the mixture design of 40SF type. Referring to the previous study by Santoyo et al., (2001) [13]; the groutability of G-class cement slurry was evaluated for the retarder–cement (r/c) ratios of 0.005, 0.010, 0.015, 0.020 and 0.025. The experimental results for the various r/c are compared in the R_m versus c chart as shown in Fig. 7. The groutability of G-class cement slurry significantly increases with the addition of a small dose of retarder. Moreover, the groutability is gradually enhanced with an increase in the r/c ratio. The optimum r/c ratio is recommended between 0.01 and 0.015 from the current experiment, because the increasing rate of groutability is greatest at the r/c ratio between 0.01 and 0.015. In addition, sufficient groutability is still maintained even 50 min after mixing the G-class cement slurry, with the r/c ratio higher than 0.015. Consequently, an addition of retarder in the G-class cement slurry mixture is an effective means to secure an appropriate groutability for geothermal wells.

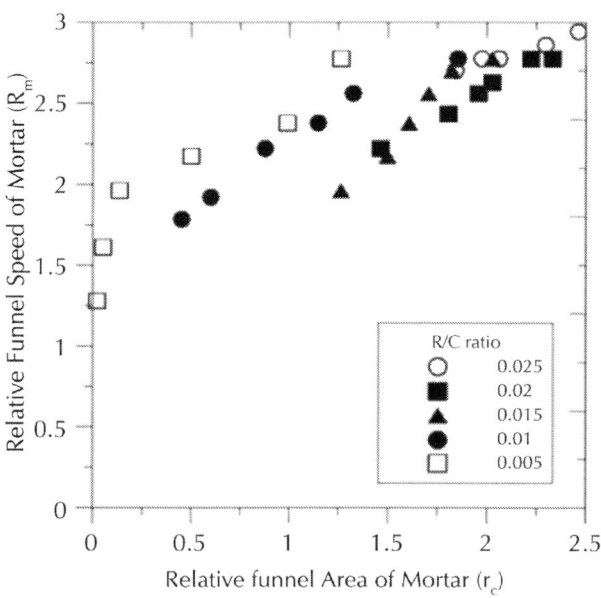

Figure 7: Groutability of G-class cement slurry for various r/c ratios with retarder.

MECHANICAL PROPERTIES OF HARDENED CEMENT

Experimental Equipment and Procedure

The uniaxial compressive strength of the hardened cement specimens was measured, to evaluate the structural integrity of geothermal well cementing that may be vulnerable to expansion/contraction of the steel casing, during operating geothermal wells. In this paper, the uniaxial compressive strength test for G-class cement specimens was performed that hardened under two different curing conditions. In order to simulate a relatively high-temperature condition in the deep subsurface, the G-class cement specimens were cured at a temperature of 100 °C in dry condition. On the other hand, the G-class cement specimens were cured in a water bath at room temperature (i.e., 21 °C) to represent a groundwater-bearing condition. The uniaxial compressive strength of G-class cement was experimentally evaluated for the four water–cement (w/c) ratios (0.55, 0.6, 0.7, 0.8) after 28 days under the previous curing conditions.

Cylindrical specimens with 100 mm in diameter and 200 mm in height were prepared for the uniaxial compression test. Three strain gages were attached on the specimen (two longitudinal directions, one lateral direction), to estimate the strain–stress relationship and the Young›s modulus of the G-class cement. The axial strain was determined by averaging the measurements from the two strain gages in the longitudinal direction. On the other hand, the strain gage installed in the lateral direction measures the lateral strain (or diametric deformation) during the test by assuring that the axial force is uniformly distributed on the cross-section of the specimen. For each curing condition and w/c ratio, four individual specimens were prepared and tested to maintain statistical reliability in determining the values of compressive strength and Young›s modulus. An automatic compression testing machine (refer to Fig. 8) was used for uniaxially loading the specimen, and the applied load and strain were measured by the load cell installed at the bottom of the specimen and the strain gages attached on the specimen, respectively. The load was applied by the stress rate of 500 kPa/s controlled by the testing machine, and the corresponding axial strain was measured.

Figure 8: Uniaxial compressive strength measurement.

Uniaxial Compression Strength and Young's Modulus

Typical stress–strain relationships for each four w/c ratio and curing condition (dry and wet) are illustrated in Fig. 9 and Fig. 10 respectively. The uniaxial compressive strength of the specimen was determined by selecting the maximum applied stress. The Young's modulus was determined by the slope from the origin to the stress level at 40% of the ultimate compressive strength and the axial strain of 5.0×10^{-5} in accordance with ASTM C469 as described in Eq. (3):

$$E = \frac{(\sigma_2 - \sigma_1)}{(\varepsilon_2 - 0.00005)}$$

(3)

where σ_1 and σ_2 are the stress corresponding to the axial strain of 5.0×10^{-5} and 40% of the ultimate compressive strength of the specimen, respectively and ε_2 is the axial strain corresponding to σ_2.

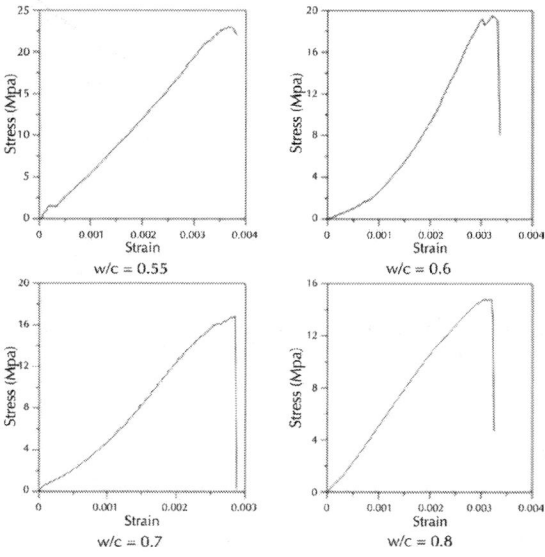

Figure 9: Stress–strain relationships for each w/c ratio cured at the dry condition.

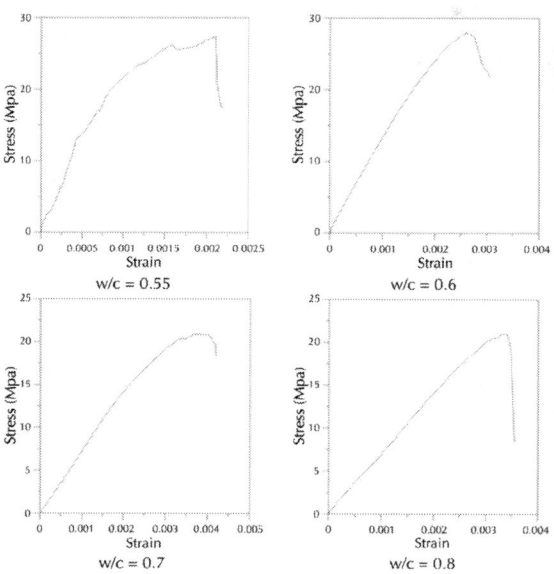

Figure 10: Stress–strain relationships for each w/c ratio cured at the wet condition.

The average uniaxial compressive strength (q) and Young's modulus (E) of G-class cement specimens are estimated for the curing conditions and w/c ratios and summarized in Table 2. The severe curing condition (i.e., dry curing at 100 °C) causes a significant reduction in the Young›s modulus by more than 50%, and a marginal decrease in the uniaxial compressive strength by about 20–25%. As can be expected, the uniaxial compressive strength and Young›s modulus decrease with an increase in the w/c ratio.

Table 2: Uniaxial compressive strength and Young's modulus of G-class cementspecimens

W/c ratio	0.55		0.6	
Curing condition	Dry[a] (100 °C)	Water[b] (21 °C)	Dry (100 °C)	Water (21 °C)
E (MPa)	6224	26,216	6379	15,183
q (Mpa)	23.04	28.97	19.50	26.23
w/c ratio	0.7		0.8	
Curing condition	Dry (100 °C)	Water (21 °C)	Dry (100 °C)	Water (21 °C)
E (MPa)	6087	15,387	5056	10,280
q (Mpa)	16.92	21.47	14.56	20.10

[a]Dry curing condition at 100 °C.

[b]Wet curing condition at 21 °C.

Philippacopoulos and Berndt (2000) [9] reported that the compressive strength of the cement specimen prepared according to the 40SF mixture design was about 38.9 ± 4.2 MPa. The cement specimen was cured at a temperature of 52 °C in wet condition, but uniaxially loaded at 200 °C. This compressive strength is slightly higher than 28.97 MPa (refer to Table 2) that is obtained for the wet curing condition at 21 °C. The difference may be attributable to the different experimental conditions (e.g., curing temperature, loading rate, etc.). In addition, Aristodimos et al. (2002) [2] recommends the minimum compressive strength of cementing for geothermal wells should be greater than 6.9 MPa. This recommendation indicates that the w/c ratios considered in this paper will not cause any structural problems with low compressive strength in the geothermal well cementing. However, in the case that the water–cement ratio is greater than 0.7, special attention should be taken to the risk of strength degradation in practice raised by severe thermal, hydraulic, and geological conditions.

THERMAL PROPERTY OF HARDENED CEMENT

Experimental Equipment and Procedure

The cement solidified in the geothermal well should have thermal conductivity sufficiently low enough to minimize heat loss during circulating up hot steam or water from the deep underground to the turbine on the ground surface. The heat loss causes a marked decline in the efficiency of electricity generation in geothermal power plants. Therefore, the thermal conductivity of cementing material should be one of the critical design factors for geothermal wells.

In this paper, the thermal conductivity of the hardened G-class cement specimens was measured using the QTM-500 thermal conductivity meter (Kyoto Electronics) that is equipped with a PD-13 probe, of which dimension is 95 mm × 40 mm (refer Fig. 11). The dimension of cement specimens cured in the mold is 50 mm × 100 mm × 50 mm, to measure the thermal conductivity. The QTM-500 thermal conductivity meter adopts the transient hot wire method [7] and [11] where the wire plays a role of both the heating element and thermometer simultaneously under the specific amount of applied power to the measurement system. The thermal conductivity (W/mK) can be mathematically calculated as follows:

$$ k = \frac{q}{4\pi(T_2 - T_1)} \ln\left(\frac{t_2}{t_1}\right) $$

(4)

where t_1 (sec) and t_2 (sec) are the two arbitrary elapse times after current flows through the probe, T_1 (°C) and T_2 (°C) are the temperature at the times t_1 and t_2, respectively and q is the applied power in the system (W).

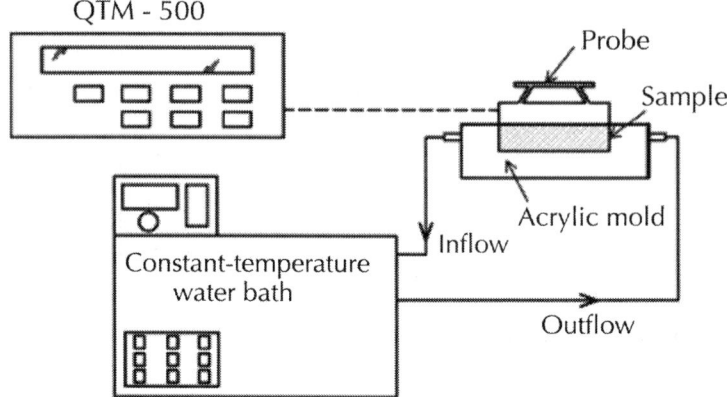

Figure 11: Schematic of thermal conductivity measurement system [8].

The thermal conductivity of G-class cement specimens was measured in two different room temperatures, i.e., 20 °C and 50 °C. In order to maintain the target temperature, a constant-temperature water bath was devised, to circulate water at a predetermined temperature, through a water jacket encasing the cement specimen (refer to Fig. 11).

Thermal Conductivity

The thermal conductivities of the G-class cement specimens measured in two different room temperatures, i.e., 20 °C and 50 °C, in consideration of the four water–cement (w/c) ratios (0.55, 0.6, 0.7, 0.8), are compared in Fig. 12. Overall, the thermal conductivity decreases with an increase in the w/c ratio for both room temperatures. This result is attributable to the fact that the smaller w/c ratio leads to a higher solid material content. The thermal conductivities measured in this paper range between 0.5 and 0.7 W/mK that are similar to those proposed by Santoyo et al. (2001) [13]. In addition, the thermal conductivity of G-class cement specimen is smaller when measured in the higher room temperature; in other words, the thermal conductivity measured at 20 °C is higher, than at 50 °C. This trend corresponds to a previous experimental result provided by Vosteen and Schellschmidt (2003) [15]. The G-class cement shows considerably low thermal conductivity, compared with the thermal conductivity of soil, usually ranging 1–2.5 W/mK [5]. This means that the G-class cement is applicable as a cementing material for geothermal wells, because it is able to prevent heat loss sufficiently, when hot water or steam is transported to the ground surface.

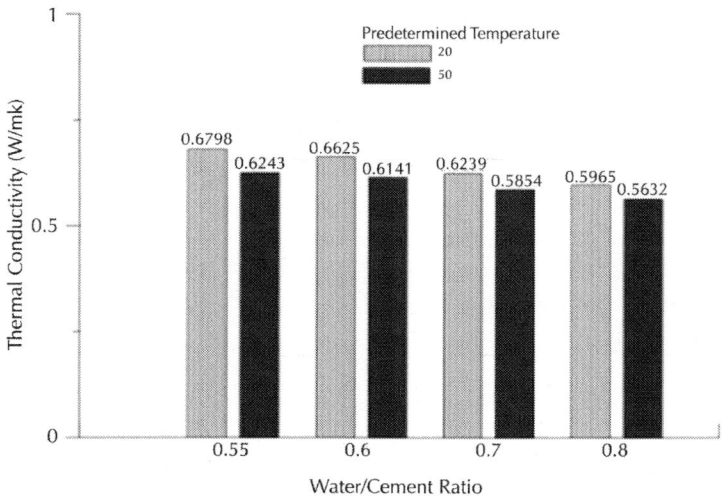

Figure 12: Comparison of thermal conductivity of cured cement specimens.

MISCELLANEOUS INDICATING EXPERIMENTS

Bleeding Potential

Non-homogeneous cementing leads to local vulnerable areas in which stress concentrates, and/or cracks may be initiated, during operating the production well. The local imperfect cementing results from bleeding due to excessive free fluid mixed in the cement slurry [12]. When exclusively considering the groutability of cement slurry with a high water–cement (w/c) ratio, a large amount of free fluid should be extracted from the cement slurry on the way of curing.

In order to quantitatively estimate the free fluid content in the cement slurry, a series of free fluid tests was performed corresponding to API specification 10A (2009) [1]. The free fluid of G-class cement was measured by pouring the cement slurry into a flask in 1 min, stirring the cement slurry in the consistometer (as shown in Fig. 13) for 20 min, and finally, extracting the free fluid formed at the slurry surface after 2 h and measuring the leftover slurry. The free fluid content (φ, %) is calculated by Eq. (5) (API specification 10 A).

$$\phi = \frac{V_{FF}\rho}{m_S} \times 100$$

(5)

where, VFF is the slurry volume leftover in the flask (cm³), ρ is the specific gravity of G-class cement (ρ = 1.91 g/cm³), and ms is the initial slurry weight (g).

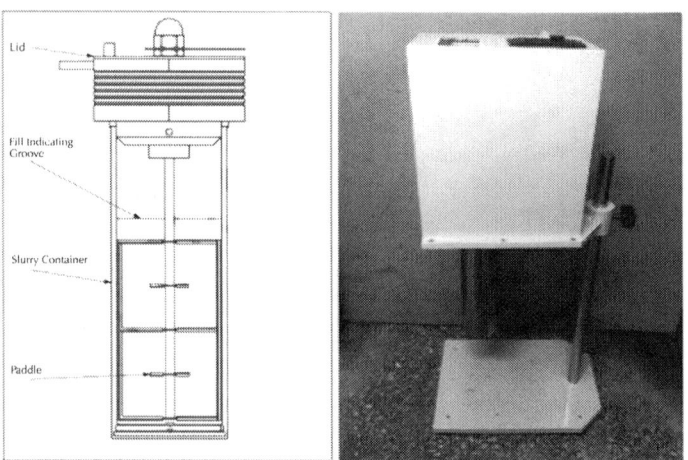

Figure 13: Schematics of consistometer.

The free fluid contents measured for the w/c ratio of 0.55, 0.6, 0.7 and 0.8 are compared in Fig. 14. Overall, the free fluid content increases, with an increase in the w/c ratio. The w/c ratio of 0.55 produces no free fluid. Even for the higher w/c ratios, the free fluid content of the G-class cement slurry shows much less than 5% that is the upper limit of free fluid content regulated by API specification 10 A. This indicates that the G-class cement used for geothermal-well cementing causes no bleeding problem.

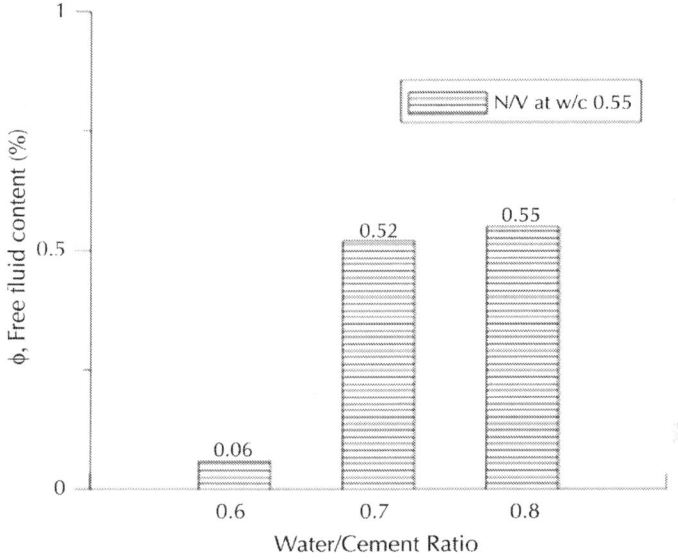

Figure 14: Free fluid content for various w/c ratios.

Phenolphthalein Indication for G-class Cement

When drilling a geothermal well, drilling mud is used to transport rock or soil debris to the ground. After drilling a borehole to the target depth, cement slurry should be injected into the annular gap between the casing and the ground formation. The pressurized cement slurry is filled from the lower part of the geothermal well, and pushes the drilling mud to be discharged to the ground. An acid–base indicator is used to judge whether cementing is completed when relatively pure cement slurry is discharged from the annular gap. The acid–base indicator responds to the pH value of the discharged material, and changes its color accordingly. The G-class cement mainly consists of bicalcium silicate ($2CaO·SiO_2$), tricalcium silicate ($3CaO·SiO_2$), tricalcium aluminate ($3CaO·Al_2O_3$) and tetracalcium aluminoferrite ($4CaO·Al_2O·3Fe_2O_3$). In addition, the hydrated G-class cement is basic with a high pH value (usually higher than 8.2) due to the existence of lime components ($Ca(OH)_2$). On the other hand, the drilling mud seems more or less neutral or slightly basic. In this paper, the pH value was measured using a pH meter, and the phenolphthalein indicator was used for observing the color change in a mixture of the G-class cement and the drilling mud. The color of phenolphthalein indicator changes to red corresponding to the pH value of 8.2–10.

Fig. 15 illustrates the pH value and the color change of phenolphthalein indicator according to various cement contents (by weight) in the mixture of G-class cement and drilling mud. The 0% of cement content means that the mixture consists of pure drilling mud. As can be expected, the pH value increases with an increase in the cement content. The phenolphthalein indicator shows no color change for the pure drilling mud (i.e., 0% of cement content) of which pH value is 7.4. When the G-class cement is more than 20%, the pH value increases rapidly to higher than 10, and the color of phenolphthalein indicator was clearly changed to red. Therefore, the phenolphthalein indicator is applicable to distinguishing the G-class cement from the drilling mud with no aid of a pH meter at construction site.

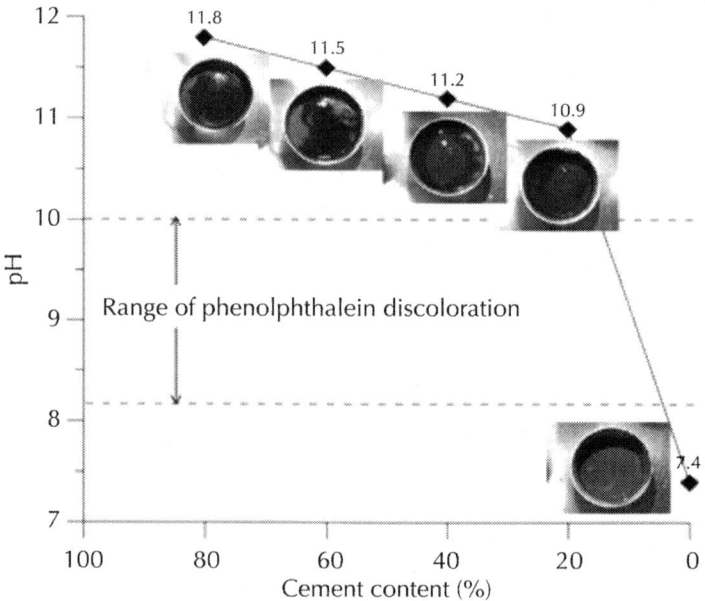

Figure 15: pH value and Phenolphthalein indication with various cement contents.

CONCLUSIONS

In this paper, significant material properties of the G-class cement through the relevant laboratory tests, and their applicability were discussed for constructing flawless geothermal power plants. The following conclusions are based on crucial findings from the experimental results.

- The groutability of the G-class cement increases with the addition of a small amount of retarder. Sufficient groutability is maintained even 50 min after mixing the G-class cement slurry, with the r/c ratio higher than 0.015. In addition, the optimum retarder–cement (r/c) ratio is recommended between 0.01 and 0.015 according to the current experimental result.

- The water–cement (w/c) ratios of 0.55, 0.6, 0.7, and 0.8, considered in this paper, will not cause any structural problems with low compressive strength in the geothermal well cementing. However, there would be a structural problem when the w/c ratio is kept extremely high in order to obtain acceptable groutability.

- The thermal conductivity of G-class cement decreases with an increase in the water–cement (w/c) ratio for both room temperatures of 20 °C and 50 °C. The overall thermal conductivity of the G-class cement is considerably low, compared with that of soil formations, which means the G-class cement is applicable as a cementing material for geothermal wells to prevent heat loss sufficiently during operating geothermal power plants.

- Even for the water–cement (w/c) ratio of the G-class cement slurry higher than 0.6, the measured free fluid contents fully satisfy the requirement regulated by API specification 10 A, which indicates the G-class cement used for geothermal-well cementing causes no bleeding problem. In addition, the phenolphthalein indicator can be used for determining whether the drilling mud in borehole is fully replaced with the G-class cement slurry instead of a pH meter.

ACKNOWLEDGMENTS

The authors appreciate the support partially by the Korea Agency for Infrastructure Technology Advancement under the Ministry of Land, Infrastructure and Transport of the Korean government (No. 13 Construction Research T01) and by National Research Foundation of Korea Government (NRF-2014R1A2A2A01007883).

REFERENCES

1.   API Specification 10A. Specification for cements and materials for well cementing. American Petroleum Institute; 2009. p. 1e5.

2.   Aristodimos J, Philippacopoulos AJ, Berndt ML. Structural analysis of geothermal well cements. Geothermics 2002;31:657e76.

3.   Edwards LM, Chilingar GV, Rieke III HH, Fertl WH. Handbook of geothermal energy. USA: Gulf Publishing Company; 1982.

4.   Felekoglu B. Utilisation of high volumes of limestone quarry wastes in concrete industry (self-compacting concrete case). Resour Conserv Recycl 2007;51(4):770e91.

5.   IGSHPA. Grouting for vertical geothermal heat pump systems: engineering design and field procedures manual. Still water: International Ground Source Heat Pump Association; 2000.

6.   JSCE, Recommendations for Design and Construction of Antiwashout Underwater Concrete, 19. Concrete library of JSCE; 1992. p. 89.

7.   Kestin J, Wakeham WA. A contribution to the theory of the transient hot-wire technique for thermal conductivity measurement. Physica 1978;92A:102e16.

8.   Lee C. Performance of ground heat exchangers for civil infrastructures Ph.D. thesis.. Seoul, South Korea: Korea University; 2012.

9.   Philippacopoulos AJ, Berndt ML. Charactization and modeling of cements for geothermal well casing remediation. Geotherm Resour Counc Trans 2000;24: 81e6.

10. DiPippo R. Geothermal power plants. In: Principles, applications, case studies and environmental impact. 2nd ed. 2008 chapter 1..

11. Roder HM. A transient hot wire thermal conductivity apparatus for fluids. J Res Natl Bureau Stand 1981;86(5):457e93.

12. Roni G, Cristiane M, Kleber T, Andre LM, Alex W. On the rheological parameter governing oilwell cement slurry stability. Annu Trans Nordic Rheol Soc 2004;12:85e91.

13. Santoyo E, Garcia A, Morales JM, Constreras E, Espinosa-paredes G. Effective thermal conductivity of Mexican geothermal cementing systems in the temperature range from 28 C to 200 C. Appl Therm Eng 2001;21:1799e812.

14. Toshifumi S. Advanced cements for geothermal wells. USA: Brookhaven National Laboratory; 2007. Report.

15. Vosteen HD, Schellschmidt R. Influence of temperature on thermal conductivity, thermal capacity and thermal diffusivity for different types of rock. Phys Chem Earth 2003;28:499e509.

16. William EG. Geothermal energy. In: Renewable energy and the environment; 2010 chapter 14..

Chapter 4

Thermal Property Measurement of Mexican Geothermal Cementing Systems Using an Experimental Technique Based on the Jaeger Method

G Espinosa-Paredes[a], García[b], E Santoyo[c],
E Contreras[b], and J.M Morales[d]

[a]UAM-Iztapalapa, Depto. de IPH, Av. San Rafael Atlixco 186, Col. Vicentina 09340, Mexico, DF, Mexico

[b]Instituto de Investigaciones Electricas, Ave, Reforma No. 113, Col. Palmira, Temixco, Mor., CP 62490, Mexico

[c]UNAM, Centro de Investigacion en Energia, Priv. Xochicalco S/N Col. Centro, Temixco, Mor., CP 62580, Mexico

[d]Centro Nacional de Investigacion y Desarrollo Tecnologico, Ave. Palmira s/n, Cuernavaca, Mor., CP 62050, Mexico

ABSTRACT

Thermal conductivity and diffusivity and specific heat capacity of six Mexican cementing systems used in geothermal well completion were determined in the temperature range from 28 to 200 °C. Thermal measurements were performed using an experimental technique based on the Jaeger method. The experimental system was calibrated by measuring the thermal properties of standard fused quartz samples. Results show that thermal conductivity and specific heat capacity depend on the particular composition of the cementing system and tend to increase with an increase in temperature, while thermal diffusivity decreases with temperature in all samples. Measurement errors were estimated less than 9.8%, 11.8% and 8.5% for the thermal conductivity, the thermal diffusivity and the specific heat capacity, respectively. Numerical correlations derived from this experimental work enable the variation of these thermal properties with temperature to be analysed.

INTRODUCTION

Thermal property measurement of geothermal cementing systems (GCS) is an important task required not only for the geothermal and oil industries but also for those researchers who are continuously compiling invaluable thermophysical property data of materials either for engineering and science handbooks or databases [1] and [2]. The knowledge of these properties at elevated temperatures is required in numerical modelling studies of a wide variety of thermal processes found in geothermal wellbore/reservoir systems[3], [4], [5], [6] and [7].

Cement thermal properties are not easy to measure accurately at high temperature due to the lack of reliable experimental instruments as well as to the high cost associated with performing laboratory measurements. Such limitations explain the current scarcity of thermophysical data and in many cases, the availability of uncertain data. Moreover, a complete determination of the thermophysical properties (thermal conductivity, thermal diffusivity and specific heat capacity) of cements used in geothermal wells is seldom found in the technical literature. In fact, very few data derived from experimental efforts have been reported under a narrow range of pressures and temperatures with uncertainties as high as ±30% [8]. Hence, a great need still exists for investigating cheaper and reliable measurement techniques for a correct assessment of these thermal properties.

In view of this lack of data, an experimental programme dealing with the determination of thermal properties of GCS was developed [9]. In the first stage of this experimental work, a measurement instrument based on the classical line source method (CLSM) was designed and operated for determining the effective thermal conductivity of six Mexican geothermal cementing samples [10]. Thermal diffusivity and specific heat capacity of these samples were not measured with the CLSM technique since it has proved to be very successful for determination of thermal conductivity only. Such a technical characteristic is due to the fact that the thermal conductivity of the CLSM does not depend on the position where temperature is measured nor on thermal contact can resistance, and the parameters needed for its determination be easily measured with high accuracy and precision. A second stage of this programme was therefore planned and executed. In this stage, a simple variant of the CLSM was implemented for putting into practice the Jaeger method (EJM: experimental Jaeger method) which enables all the thermal properties of cements to be simultaneously measured. The aim of the present paper is to describe the theoretical and technical aspects of this experimental technique as well as to show the results obtained in the thermophysical evaluation of the same Mexican geothermal cementing samples used in the study describe in Refs. [9] and [10].

THEORETICAL ASPECTS OF THE EXPERIMENTAL JAEGER METHOD

The experimental technique, EJM is based on the concept of an infinite line heat conductive medium with an ideal heat source immersed in it. The line source dissipates heat uniformly at a constant rate along its axial length. Such theoretical concepts have been described in detail by Santoyo et al. [10]. The basic difference between the CLSM and the EJM method is that the former is only recommended for measuring thermal conductivity (k) using the temperature history logged close to the heat source, while the EJM, is primarily used for the determination of the thermal diffusivity (α) from the temperature data measured at a known distance from the heat source (at $r=a$, Fig. 1). These data are later extended in this method to the calculation of both the thermal conductivity (k) and the specific heat capacity (Cp) of the sample.

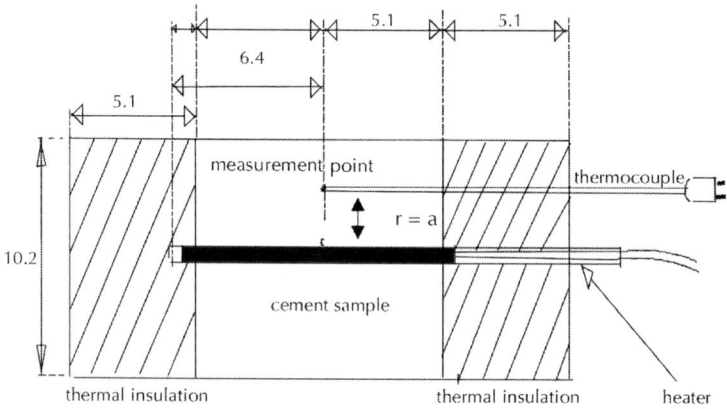

Figure 1: Instrumented sample for thermal property measurement by Jaeger's method (units are in cm).

Mathematical Model of the Experimental Jaeger Method

The thermal perturbation near an ideal linear heat source of constant intensity immersed in an infinite medium has a theoretical analytical solution given by Eq. (1) which provides the temperature field near a line heat source [10], [11] and [12]:

$$T(a,t) = -\frac{Q}{4\pi k}E_1\left(\frac{-a^2}{4\alpha t}\right)$$

(1)

where Q is the heat input per unit length; k is thermal conductivity; is thermal diffusivity; $E_1(x)$ is the exponential integral which is listed in mathematical tables; $T(a,t)$ is the temperature rise at time t and at a point located at a distance $r=a$ from the linear heat source; and a is the observation point for temperature measurement (see Fig. 1). Using Eq. (1) at times $2t$ and t yields

$$\frac{T(a, 2t)}{T(a, t)} = \frac{E_1\left(\dfrac{a^2}{8\alpha t}\right)}{E_1\left(\dfrac{a^2}{4\alpha t}\right)}$$

(2)

The right hand side of Eq. (2) has the form

$$G(x) = \frac{E_1(x)}{E_1(2x)}$$

(3)

where

$$x = \frac{a^2}{8\alpha t}$$

(4)

and constitute the mathematical model proposed by Jaeger for a simultaneous determination of thermal diffusivity and conductivity [11] and [12]. The practical application of EJM is rather simple from a conceptual point of view. Different values of the $T(a,2t)/T(a,t)$ ratio for different values of time t are determined from the experimental temperature–time record. Since these values are governed by Eq. (2), therefore, the argument $(x=a^2/8\alpha t)$ can be calculated for each value of the ratio $T(a,2t)/T(a,t)$ using the theoretical function given by Eq. (3). This is possible because the $G(x)$ function is single valued. This implies that for each value of the argument x there is a corresponding unique value of the $G(x)$ function and vice versa. The $G(x)$ function can easily be obtained from both first order exponential integral equations and tabulated values. Once the argument x is determined, the thermal diffusivity (α) of a given specimen can be computed from Eq. (4). Knowledge of the temperature–time history at a given point enables the thermal conductivity (k) of the sample material to be later estimated by use of the following equation:

$$k = \frac{Q}{4\pi T(a, t)} E_1\left(\frac{a^2}{4\alpha t}\right)$$

(5)

Considering that $(a^2/4\,t)=2x$, then

$$k = \frac{Q}{4\pi T(a, t)} E_1(2x)$$

(6)

From a conceptual point of view, the determination of both thermal diffusivity and conductivity requires only that the procedure described above be applied for any single value of the relation $T(a,2t)/T(a,t)$, and the values of these properties obtained in this way must be independent of the time t considered. Likewise in the measurement of thermal diffusivity the line source strength does not interfere whereas in the determination of the conductivity, the distance between the position of the heat source and the point where temperature is measured does not intervene in its computation. In addition, only first order exponential integral data tables for the determination of the $G(x)$ function value, which correspond to the $(2x)$ argument, are required for the computation of thermal conductivity (Eq. (1)).

The specific heat capacity of the sample specimen can finally be calculated from the definition of thermal diffusivity using the k and α experimental values, together with the cement density (ρ), i.e.,

$$\alpha = \frac{k}{\rho c}$$

(7)

EXPERIMENTAL SET-UP

The experimental system used for thermal property measurement was fundamentally the same instrument used with the CLSM technique [10]. According to the line heat source theory, the designed system must be able to supply the constant strength of the heat source and allow for transient temperature measurement at two positions. One such position is close to the line source (used by the CLSM: Ref. [10]), and the other one at a known distance $r=a$ from the source (for use in the EJM, Fig. 1). Heating and data acquisition hardware systems were coupled for obtaining an appropriated configuration of the measurement instrument (see Fig. 2 reported by Santoyo et al. [10]). A very brief sketch of this equipment is presented in the following sections. A further description of the technical design aspects of each one of their components was given in the first stage of this experimental measurement programme (see Ref. [10]).

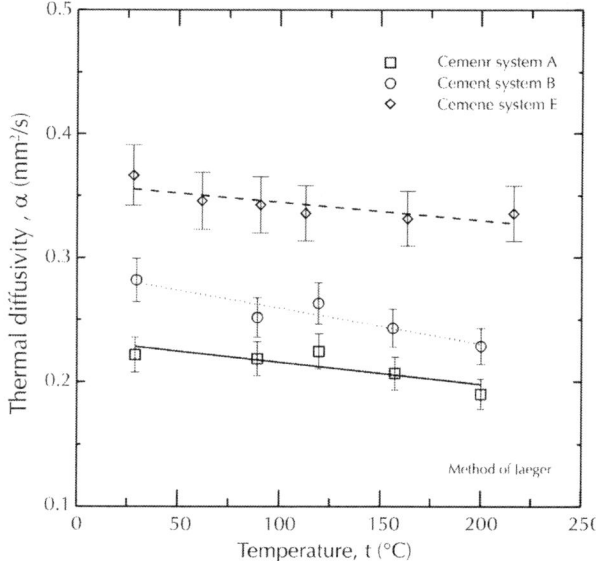

Figure 2: Thermal diffusivity results obtained by the EJM technique for geothermal cement systems A, B and E..

Heating System

This heating module enabled a thermal disturbance to be applied to the sample specimen under study. With this purpose, it provided a constant heat rate by applying a constant electric current through the heater which was immersed within the specimen. The main components of this module were an electrical resistance heater or linear heat source, a direct current source, a voltmeter and an ammeter.

Data Acquisition System

When using the EJM technique implemented in the experimental set-up, the data acquisition system (DAS) allowed for transient measurement of the sample temperature rise at a point located at a known distance from the heat source ($r=a$, Fig. 1). DAS consisted of a data acquisition unit, an A/D converter, a signal amplifier, a voltage compensatory device, several thermocouples, a 0 °C thermal reference unit and a furnace.

Temperature Measuring and Recording System Resolution

A reading of 5 V was typically achieved with a base temperature of 25 °C and an amplification of 5000× gain. A resolution of 0.002 V in the A/D converter produced a resolution of 0.01 °C in the temperature recording system which is sufficient for the purposes of the EJM technique.

Sample Preparation and Instrumentation Procedures

Solid cement samples 10.16 cm diameter and 10.16 cm length obtained from cement slurries were carefully prepared in moulds using the cement specimen casting procedure reported by Santoyo et al. [10]. The cement slurries were previously prepared in accordance with the American Petroleum Institute (API) specifications [13]. The cement specimens were then cured for one week at 80 °C and 20 MPa in a high pressure-curing chamber. Prior to solidification, the samples were instrumented with the heater at the sample longitudinal axis and a thermocouple inside the sample, at a known distance from the heater (Fig. 1). As mentioned earlier, for each cementing system, two instrumented samples were prepared. The first sample was used for measuring its effective thermal conductivity by the CLSM technique [10]. The second sample was primarily used with the EJM technique for determining the thermal diffusivity, thermal conductivity and specific heat capacity which is the fundamental part of this work.

Experimental Procedure

The experimental procedure used to record the temperature and time data consisted of applying an electrical current of constant intensity to the sample heater during a time period. This period usually lasted from 3 to 5 min for logging both temperature and time data at a frequency of three readings per second. During this time, the voltage drop across the heater and the circulating current were also recorded, such that the dissipated power per unit length could be computed to complete all the experimental data required by the EJM. For reaching temperatures above ambient, the samples were uniformly heated in the muffle to the next temperature step. The sample was maintained at the new constant temperature for applying electrical energy to the heater and to record all the transient experimental data at this new temperature level and so on until completing all the desired temperature steps for thermal property measurement. The heating of the sample, the temperature stabilisation and the final thermal property measurement, at the various temperature steps, was therefore a time consuming task since these activities required several days.

Data Reduction of the Experimental Jaeger Method

The data reduction process was carried out as follows:

- From the experimental temperature and time data logged, $T(nt_0)$ was calculated for values of n=1,2,3,...(=t_T/t_0) where t_T is the total data recording time and t_0 is a reference time whose value depends on the value of t_T and the data recording frequency. A value of 2–4 s was adequate for a frequency of 60 readings per min and a total recording time between 3 and 5 min.

- Values of the ratio $T(2nt_0)/T(nt_0)$ were then estimated from n=1,2,3,...,n_{max}/2.

- The argument x corresponding to each $T(2nt_0)/T(nt_0)$ ratio value of the function $G(x)$ was next computed either with the aid of a graph or by the equations defining the exponential integral to obtain the $G(x)$ function.

- The values of the $(a^2/8ant_0)$ group were obtained by Eq. (4) and the values of x were then multiplied by the corresponding value of n to obtain $(a^2/8at_0)$.

- The calculations generated in step 4 were analysed to identify the interval where the values of the group $(a^2/8ant_0)$ remain constant within the random fluctuations of the experiment. The average value of this group was obtained in the interval of relative invariability and from this mean value, the thermal diffusivity was obtained using the value of the thermocouple position (a) and t_0.

- The values the group $(a^2/8ant_0)$ were then multiplied by a factor of 2 to generate the group of values $(a^2/4at_0)$.

- With the values of the argument $(a^2/4\alpha t_0)$, the exponential integral E_1 was computed.
- The values of the ratio $[T(nt_0)/E_1(a^2/4\alpha t_0)]$ were then calculated using Eq. (1), i.e., $Q/(4\varpi k)$ was computed by dividing each value of $T(nt_0)$, from $n=1$ to $n=n_{max}/2$, by the corresponding value of $E_1(a^2/4\alpha t_0)$ for the same n. Then, the $Q/(4\varpi k)$ values were analysed to find the interval where the ratio $[T(nt_0)/E_1(a^2/4\alpha nt_0)]$ remains constant within experimental fluctuation. The average value was computed and assigned to the group $Q/(4\varpi k)$, and Eq. (6) was solved for the thermal conductivity, k.

The specific heat was finally computed from Eq. (7).

SYSTEM CALIBRATION

A calibration procedure of the experimental system was accomplished by performing two series of standard tests where the technique uncertainty and repeatability were evaluated. Fused quartz was used as standard because its thermal properties are well characterised. Two samples of this standard material were separately examined. Twenty-one independent experimental measurements were performed. In all calibration runs, the experimental conditions and parameters were kept constant (e.g. sample heating time, frequency of data acquisition, current intensity applied to the heater, etc.).

In the first series of experiments, the mean values of thermal diffusivity (α), thermal conductivity (k) and specific heat capacity (Cp) were estimated as 0.824 mm^2 s^{-1} (±0.051); 1.383 W m^{-1} K^{-1} (±0.065) and 0.758 kJ kg^{-1} K^{-1} (±0.034), respectively (Table 1). These experimental results compare well with data reported for the standard sample: 0.825 mm^2 s^{-1} (±0.025); 1.3794 W m^{-1} K^{-1} (±0.024) and 0.753 kJ kg^{-1} K^{-1} [9] and [14]. Such a comparison yields accuracy errors less than 0.12%, 0.26% and 0.66% for α, k and Cp, respectively. In the second calibration series, mean values of α, k and Cp were estimated as 0.822 mm^2 s^{-1} (±0.054); 1.389 W m^{-1} K^{-1} (±0.026) and 0.773 kJ kg^{-1} K^{-1} (±0.047), respectively (Table 2). These calibration data also show a good agreement with the standard data reported for fused quartz samples[9] and [14].

Table 1: Thermal property results obtained by the EJM technique for fused quartz samples in the first series of calibration tests

Test	Thermal diffusivity (mm² s⁻¹)	Thermal conductivity (W m⁻¹ K⁻¹)	Specific heat capacity (J kg⁻¹ K⁻¹)
cj-04	0.7440	1.2920	0.7893
cj-10	0.8533	1.3359	0.7116
cj-11	0.8401	1.3603	0.7360
cj-12	0.9132	1.3882	0.6914
cj-13	0.8551	1.3674	0.7314
cj-14	0.8799	1.4018	0.7261
cj-15	0.8000	1.4552	0.7923
cj-16	0.7647	1.3050	0.7790
cj-17	0.7566	1.3137	0.7666
cj-18	0.8378	1.4845	0.7549
cj-19	0.8384	1.4676	0.7958
cj-20	0.8065	1.4230	0.7973
Average	0.8241	1.3829	0.7581
Standard deviation	0.0513	0.0652	0.0342

Table 2: Thermal property results obtained by the EJM technique for fused quartz samples in the second series of calibration tests

Test	Thermal diffusivity (mm² s⁻¹)	Thermal conductivity (W m⁻¹ K⁻¹)	Specific heat capacity (J kg⁻¹ K⁻¹)
cj-21	0.8216	1.4214	0.6728
cj-22	0.8547	1.3839	0.7358
cj-23	0.8664	1.3721	0.7199
cj-24	0.8332	1.4061	0.7667
cj-25	0.8332	1.3738	0.7678
cj-26	0.7865	1.4091	0.8082
cj-27	0.7777	1.3568	0.7939

cj-28	0.7646	1.4294	0.8493
cj-29	0.7953	1.3491	0.7711
cj-30	0.7931	1.4133	0.8100
cj-31	0.7732	1.3726	0.8070
Average	0.8216	1.3898	0.7729
Standard deviation	0.0545	0.02594	0.04704

RESULTS AND DISCUSSION

Six GCS were again selected and used for applying the experimental EJM technique. Such samples represent the compositions commonly used in the Mexican geothermal well industry. The compositions of the GCS have already been reported by Santoyo et al. (see Ref. [10, Table 2]).

Thermal Diffusivity of Cementing Systems

Thermal diffusivity of the six geothermal cement systems (GCSs) was determined using EJM method. The experimental results obtained for the A, B and E GCSs are shown in Fig. 2, while the corresponding results for the C, D and F GCSs are presented in Fig. 3. The best fitting straight lines are also shown for each set of data. It may be observed that in all cases, thermal diffusivity decreases with increasing temperature. From a conceptual point of view, thermal diffusivity is independent of the line source strength Q but depends on the square of the radial distance from the source where the temperature is measured. However, in spite of the errors associated with the data points of Fig. 3 and Fig. 4, well-defined trends of thermal diffusivity with temperature are observed. The diffusivity curves for the D and F GCS are very close and this may be due to their chemical compositions where the weight fraction of API cement G, silica fluor and water are similar for both systems (see Ref. [10, Table 2]). The order of magnitude of the thermal diffusivity values obtained in the present study are in line with reported values for other cementing systems at ambient or near ambient conditions [9], [15] and [16].

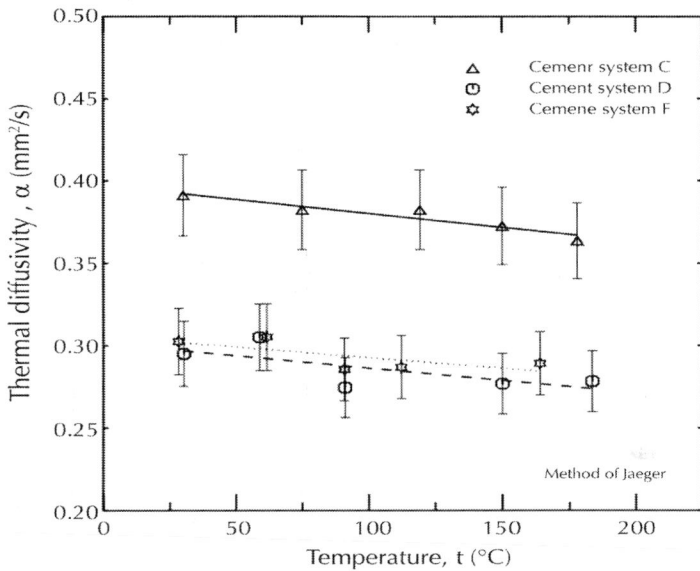

Figure 3: Thermal diffusivity results obtained by the EJM technique for geothermal cement systems C, D and F.

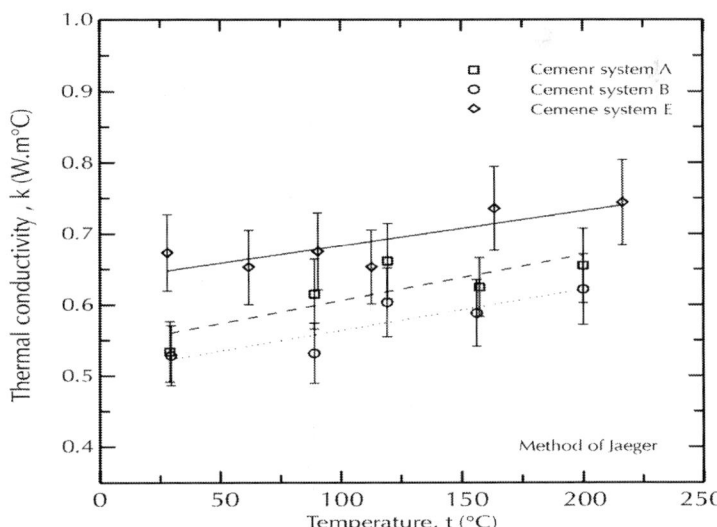

Figure 4: Thermal conductivity results obtained by the EJM technique for geothermal cement systems A, B and E..

Thermal Conductivity of Geothermal Cementing Systems

The thermal conductivity and temperature measurements of the six GCS were also divided in two groups. The experimental measurements related to the first group (A, B and E) are shown in Fig. 4, while the second group (C, D and F) are presented in Fig. 5. Experimental uncertainties of the thermal conductivity are represented as statistical error bars. Measurement errors due to temperature readings were neglected since these were very small. The best fitting straight lines are also represented for each set of data. It may be observed from these figures, that the thermal conductivity results show an increasing variation with increasing temperature for all GCSs. The magnitude of the experimental errors observed in these samples are greatly attributed to the chemical and mineralogical composition of the cement samples [10]. However, the magnitude of these errors may also be due to the radial position from the heat source where the temperature is measured since it appears to the second power in the argument of the exponential integral (see Eq. (5)). Thus, from an experimental point of view, the application of EJM method is sensitive to errors relating the position of the temperature-sensing device.

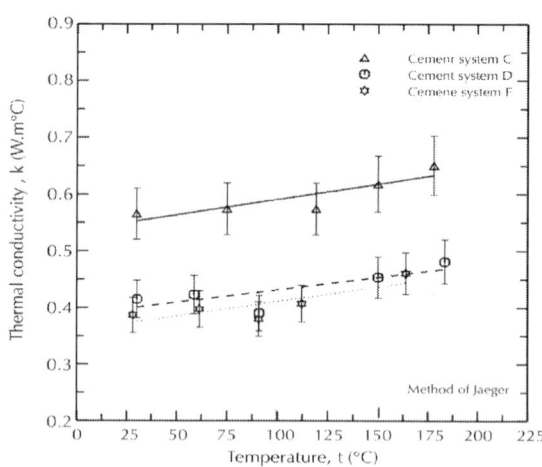

Figure 5: Thermal conductivity results obtained by the EJM technique for geothermal cement systems C, D and F.

Specific Heat Capacity of Set Cementing System

The specific heat capacity results of the cement systems considered in the present study are shown in Fig. 6 and Fig. 7. These results were obtained via Eq. (7) using the thermal diffusivity and conductivity results as well as the cement densities. The best fitting straight lines are also shown for each set of data. It is clearly observed from these figures that specific heat capacities increase with increasing temperature for all cases. These results are of the same order of magnitude as those of similar materials like rocks, bricks, concrete and cements at ambient conditions [9].

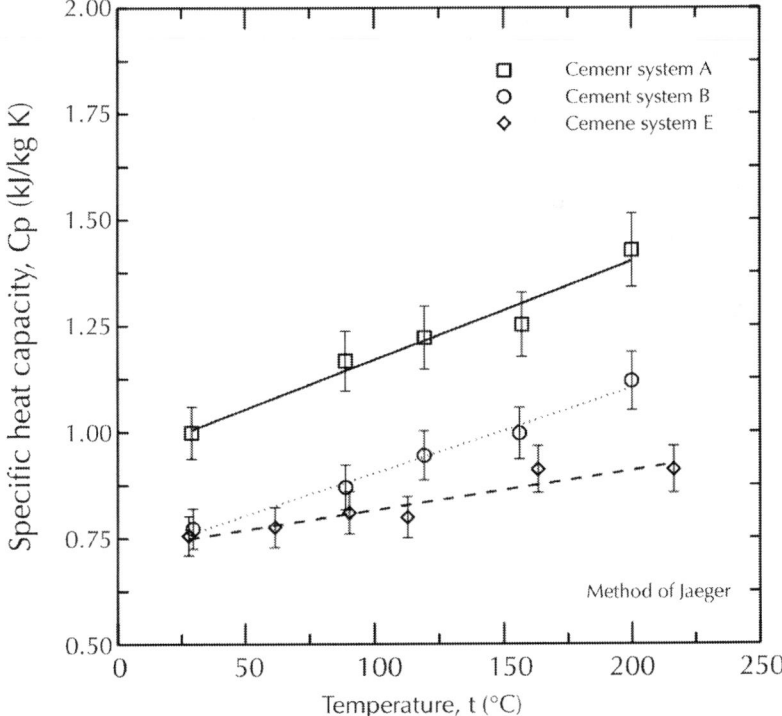

Figure 6: Specific heat capacity results obtained by the EJM technique for geothermal cement systems A, B and E..

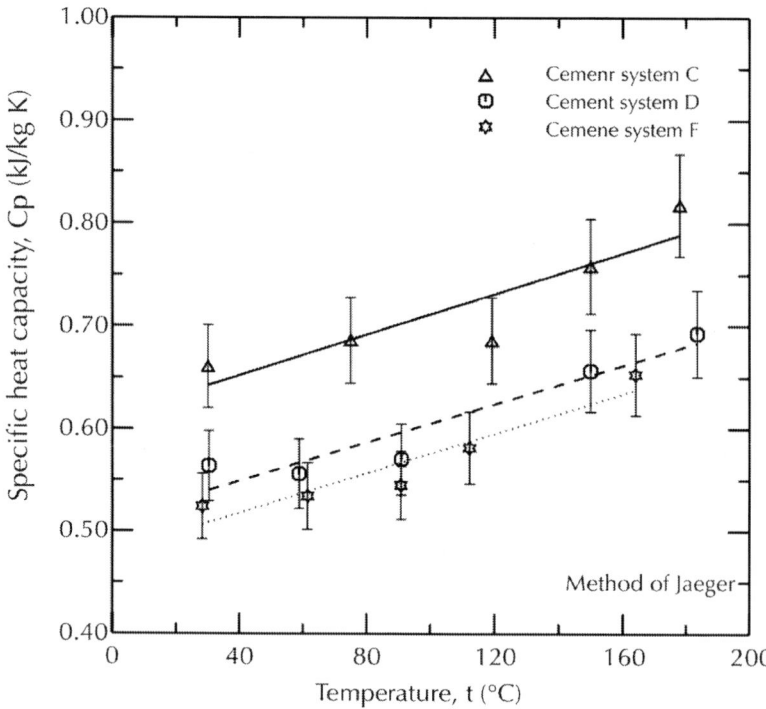

Figure 7: Specific heat capacity results obtained by the EJM technique for geothermal cement systems C, D and F.

Regression Analysis

A compilation of the thermal property results obtained with the use of the EJM technique, including their experimental errors are given in Table 3. A statistical data reduction procedure for correlating the thermal conductivity and diffusivity and specific heat capacity (k, α and C_p) with temperature (T) was performed. Such a statistical procedure was carried out to obtain numerical correlations and their associated errors. A comprehensive statistical methodology based on weighted linear regression models with errors was used. Such a methodology uses a fitting straight line procedure when both variables (X–Y) are subject to error. Thus, values of the intercept (A) and the slope (B) as well as their corresponding standard errors: σA and σB are computed [17] and [18].

Table 3: Compilation of thermal property measurements of the Mexican geothermal cement systems obtained by use of the EJM technique

Temperature ($\pm\sigma_t$) (°C)	k ($\pm\sigma$k) (W m^{-1} K^{-1})	a ($\pm\sigma$a) ($mm^2\,s^{-1}$)	ρ ($\pm\sigma$ρ) (kg m^{-3})	Cp ($\pm\sigma$Cp) (kJ kg^{-1} K^{-1})
Cement system—A:				
29.0 (0.1)	0.5338 (0.0427)	0.22168 (0.01470)	2411.6 (24)	0.9985 (0.0607)
89.0 (0.3)	0.6551 (0.0492)	0.21855 (0.01449)		1.1671 (0.0709)
119.0 (0.4)	0.6616 (0.0529)	0.22447 (0.01488)		1.2221 (0.0743)
156.0 (0.5)	0.6250 (0.0500)	0.20691 (0.01372)		1.2526 (0.0761)
200.0 (0.7)	0.6551 (0.0524)	0.19027 (0.01262)		1.4277 (0.0868)
Cement system—B:				
29.5 (0.1)	0.5284 (0.0423)	0.28171 (0.01868)	2427.5 (24)	0.7727 (0.0469)
89.0 (0.3)	0.5316 (0.0425)	0.25172 (0.01669)		0.8700 (0.0529)
119.3 (0.4)	0.6035 (0.0482)	0.26304 (0.01744)		0.9452 (0.0575)
156.2 (0.5)	0.5883 (0.0470)	0.24309 (0.01612)		0.9970 (0.0606)
200.0 (0.7)	0.6217 (0.0497)	0.22869 (0.01516)		1.1199 (0.0681)
Cement system—C:				
30.0 (0.1)	0.5656 (0.0452)	0.39120 (0.02594)	2188.2 (22)	0.6607 (0.0402)
75.0 (0.3)	0.5743 (0.0459)	0.38238 (0.02535)		0.6863 (0.0417)
119.2 (0.4)	0.5740 (0.0459)	0.38238 (0.02535)		0.6859 (0.0417)
150.0 (0.5)	0.6181 (0.0494)	0.37254 (0.02470)		0.7583 (0.0461)
178.0 (0.6)	0.6508 (0.0520)	0.36364 (0.02411)		0.8179 (0.0497)
Cement system—D:				
30.3 (0.1)	0.4149 (0.0331)	0.29502 (0.01956)	2494.0 (25)	0.5639 (0.0343)
58.8 (0.2)	0.4230 (0.0338)	0.30500 (0.02022)		0.5561 (0.0338)
91.0 (0.3)	0.3905 (0.0312)	0.27464 (0.01821)		0.5702 (0.0347)
150.0 (0.5)	0.4534 (0.0363)	0.27685 (0.01836)		0.6566 (0.0399)
183.5 (0.6)	0.4810 (0.0384)	0.27837 (0.01846)		0.6929 (0.0421)
Cement system—E:				
27.9 (0.1)	0.6732 (0.0539)	0.36618 (0.02428)	2433.0 (24)	0.7557 (0.0459)
61.7 (0.2)	0.6530 (0.0522)	0.34582 (0.02293)		0.7761 (0.0472)
90.5 (0.3)	0.6752 (0.0540)	0.34256 (0.02271)		0.8101 (0.0492)
112.8 (0.4)	0.6534 (0.0523)	0.33587 (0.02227)		0.7996 (0.0486)
163.5 (0.6)	0.7357 (0.0588)	0.33155 (0.02198)		0.9120 (0.0555)
216.3 (0.7)	0.7442 (0.0595)	0.33555 (0.02225)		0.9116 (0.0554)
Cement system—F:				
28.2 (0.1)	0.3867 (0.0309)	0.30243 (0.02005)	2439.0 (24)	0.5243 (0.0318)
61.5 (0.2)	0.3975 (0.0318)	0.30505 (0.02022)		0.5343 (0.0325)
90.8 (0.3)	0.3796 (0.0304)	0.28557 (0.01893)		0.5450 (0.0331)
112.2 (0.4)	0.4069 (0.0325)	0.28691 (0.01902)		0.5815 (0.0353)
164.0 (0.6)	0.4606 (0.0368)	0.28916 (0.01917)		0.6530 (0.0397)

The regression procedure was used to determine all the statistical parameters required for evaluating the existence of linear correlations between α, k and Cp (dependent variable: Y) and T (independent variable: X) in the set of experimental data of each GCS. The linear correlation coefficient (R) and the correlation probability $P(R,N)$ were also obtained according to the statistical correlation theory proposed by Bevington[19]. The interpretation of high values of $P(R,N)$ would indicate that both variables Y (α, k or Cp) and X (T) are non-correlated while lower values of it would confirm that such variables are appropriately correlated with a good confidence level.

Table 4 summarises the complete statistical evaluation of the temperature dependency of the thermal property for all the measurements performed with the EJM technique. These linear correlations were also plotted for representing the best-fit line of each series of experimental data (Fig. 2, Fig. 3, Fig. 4, Fig. 5, Fig. 6 and Fig. 7). All experimental measurements show linear correlation coefficients (R) ranging from 0.7098 to 0.9936. The lowest correlation coefficient was related to the thermal diffusivity–temperature equation of the D cementing system ($R=0.7098$). Even though, such a correlation value could be considered statistically low, the correlation probability $P(R,N)$ indicates that and T are correlated variables with an approximated confidence level of 95%. On the other hand, application of the linear correlations obtained from results of the EJM technique (Table 4) for predicting the thermal diffusivity, thermal conductivity and specific heat capacity of GCSs over the temperature range shown in Fig. 2, Fig. 3, Fig. 4, Fig. 5, Fig. 6 and Fig. 7 should produce average propagated errors ranging from 7.3% to 9.8% for thermal diffusivity; from 9.0% to 11.8% for thermal conductivity; and from 5.82% to 8.5% for specific heat capacity.

Table 4: Statistics results of the temperature dependency on the thermal property (k, a, Cp) measurements carried out in the Mexican geothermal cement samples by mean of the EJM technique

GCS/ thermal property	Number of data (n)	Intercept-A($\pm\sigma$A)	Slope-B ($\pm\sigma$B)	Linear correlation coefficient (R)	Probability of correlationP(R,N)*
Thermal conductivity (k)	5	0.53405 (0.0457)	−0.0006894 (0.000361)	0.8227	0.0872
Thermal diffusivity (a)	5	0.23423 (0.0146)	−0.0001875 (0.000105)	0.8227	0.0872
Heat specific capacity (Cp)	5	0.93790 (0.0660)	0.0023060 (0.000550)	0.9799	0.0034
B					
Thermal conductivity (k)	5	0.50492 (0.0442)	0.0005745 (0.000345)	0.8843	0.0464
Thermal diffusivity (a)	5	0.28744 (0.0181)	−0.0002872 (0.000130)	0.9328	0.0207
Heat specific capacity (Cp)	5	0.70730 (0.0510)	0.0019630 (0.000430)	0.9936	0.0006
C					
Thermal conductivity (k)	5	0.53811 (0.0478)	0.0005213 (0.000406)	0.8788	0.0499
Thermal diffusivity (a)	5	0.39733 (0.0265)	−0.0001716 (0.000213)	0.9468	0.0146
Heat specific capacity (Cp)	5	0.61690 (0.0430)	0.0009320 (0.000370)	0.9003	0.0372
D					
Thermal conductivity (k)	5	0.38717 (0.0313)	0.0004235 (0.000282)	0.7949	0.1079
Thermal diffusivity (a)	5	0.30274 (0.0180)	−0.0001439 (0.000150)	0.7098	0.1792
Heat specific capacity (Cp)	5	0.51310 (0.0320)	0.0009120 (0.000300)	0.9507	0.0130
E					

Thermal conductivity (k)	6	0.63424 (0.0454)	0.0004778 (0.000370)	0.8301	0.0408
Thermal diffusivity (a)	6	0.35835 (0.0195)	−0.0001399 (0.000149)	0.8036	0.0541
Heat specific capacity (Cp)	6	0.72270 (0.0400)	0.0009260 (0.000330)	0.9467	0.0042
F					
Thermal conductivity (k)	5	0.36141 (0.0319)	0.0004764 (0.000329)	0.8279	0.0834
Thermal diffusivity (a)	5	0.35090 (0.0197)	−0.0001258 (0.000190)	0.7174	0.1724
Heat specific capacity (Cp)	5	0.48280 (0.0330)	0.0009190 (0.000350)	0.9490	0.0137

CONCLUSIONS

The effect of temperature on the thermal conductivity and diffusivity and specific heat of six GCSs commonly used in Mexican geothermal well industry was experimentally studied in the range from 28 to 220 °C using the EJM technique. Thermal diffusivity of all GCS was found to decrease with increasing temperature and its values are of similar magnitude as those of similar materials. Thermal conductivity increases with temperature for all cement systems. Cement mineralogical composition was clearly identified to play an important role in the thermal conductivity values obtained for these cements. A comprehensive study for explaining the differences observed in the calculation of the thermal conductivity of such cementing materials by use of both the CLSM and the EJM techniques is being carried out and it will be published further on.

On the other hand, the specific heat capacity of all cement systems was found to increase with increasing temperature. This behaviour has also been observed in similar and thermal insulating materials like expanded perlite.

In principle, the experimental results of the present study can reliably be used to compute temperatures in geothermal wells during drilling, circulation and shut-in by numerical simulation where the effect of temperature-dependent thermal properties of the cement systems is accounted for in calculating the well heat transfer coefficients. Finally, it is very important to point out that the present experimental technique can easily be applied to the determination of thermophysical properties not only to cementing samples but also to any solid materials (rocks, concrete or polymers).

ACKNOWLEDGEMENTS

The authors wish to thank to the late Professor F.A. Holland for his notable contribution to the development of geothermal research in Mexico. Additional thanks are due to Professor Ian D. Cluckie of the Telford Research Institute, U. Salford, UK (now at University of Bristol, UK) for his support. The authors also appreciate the support from Instituto de Investigaciones Eléctricas and the comments and suggestions from Dr. David A. Reay (Editor) which contributed considerably in improving the manuscript.

REFERENCES

1.   R.H. Perry, C.H. Chilton, Chemical Engineers' Handbook, McGraw-Hill, New York, 1997.

2.   D.R. Lide, H.V. Kheiaian, CRC Handbook of Thermophysical and Thermochemical Data, CRC Press, Boca Raton, FL, 1997.

3.   G. Garcia-Estrada, A. Lopez-Hernandez, R.M. Prol-Ledesma, Temperature–depth relationships based on log data from the Los Azufres geothermal field, Mexico, Geothermics 30 (2001) 111–132.

4.   S.H. Bittleston, A two-dimensional simulator to predict circulating temperatures during cementing operations, Proceedings of 65th Annual Technical Conference and Exhibition, Society of Petroleum Engineers, New Orleans, LA, USA, September 23–26, 1990, pp. 443–454.

5.   G. Bjornsson, G. Bodvarsson, A survey of geothermal reservoir properties, Geothermics 19 (1990) 17–27.

6. R.M. Beirute, A circulating and shut-in well-temperature-profile simulator, Journal of Petroleum Technology September (1991) 1140–1146.

7. A. Garcia, E. Santoyo, G. Espinosa, I. Hernandez, H. Gutierrez, Estimation of temperatures in geothermal wells during circulation and shut-in in the presence of lost circulation, Transport in Porous Media 33 (1998) 103–127.

8. M.A. Goodman, Arctic well completion series—Part 3: Here's what to consider when cementing permafrost, World Oil 177 (1977) 81–90.

9. J.M. Morales, Thermal property measurement of cement systems used in geothermal wells, in: M.Sc. Thesis, University of Salford, UK, 1997.

10. E. Santoyo, A. Garcia, J.M. Morales, E. Contreras, G. Espinosa-Paredes, Effective thermal conductivity of Mexican geothermal cementing systems in the temperature range from 28 to 200 C, Applied Thermal Engineering 21 (2001) 1799–1812.

11. H.S. Carslaw, J.C. Jaeger, Conduction of Heat in Solids, Oxford University Press, London, 1959.

12. J.C. Jaeger, The use of complete temperature–time curves for determination of the thermal conductivity with particular reference to rocks, Australian Journal of Physics 12 (3) (1959) 203–217.

13. Specification 10: Specification for Materials and Testing for Well Cements, fifth ed., American Petroleum Institute, Washington, DC, USA, 1990 (pp. 9–17).

14. Fused Quartz Catalogue of Products, US Fused Quartz Company Inc., Fairfield, NJ, USA, 1997 (p. 38).

15. L.R. Raymond, Temperature distribution in a circulating drilling fluid, Journal of Petroleum Technology March (1969) 333–341.

16. F.C. Arnold, Temperature variation in a circulating wellbore fluid, Journal of Energy Resources Technology 112 (1990) 79–83.

17. K.I. Mahon, The New York regression: application of an improved statistical method to geochemistry, International Geology Review 38 (1996) 293–303.

18. D. York, Least squares fitting of a straight line with correlated errors, Earth and Planetary Science Letters 5 (1969) 320–344.

19.   P.R. Bevington, Data Reduction and Error Analysis for the Physical Sciences, McGraw-Hill, New York, 1969.

Chapter 5

Cementing Mechanism of Potassium Phosphate Based Magnesium Phosphate Cement

Zhu Ding[a], Biqin Dong[a], Feng Xing[a], Ningxu Han[a], and Zongjin Li[b]

[a]School of Civil Engineering, Guangdong Province Key Laboratory of Durability for Marine Civil Engineering, Shenzhen University, Shenzhen 518060, PR China
[b]Department of Civil Engineering, The Hong Kong University of Science and Technology, Clear Water Bay, Kowloon, Hong Kong, China

ABSTRACT

Magnesium phosphate cements (MPCs) are materials that belong to chemically bonded ceramic materials. They have a wide range of potential applications, due to their superior performance. In this paper, the reaction products and cementing mechanism of magnesium phosphate bonded cement based on the dead burned magnesia and the mono-potassium phosphate (MPP) are investigated. Fine powder and grains of dead burned magnesia were used to prepare pure cement paste and bonding cluster samples, respectively. The cement reaction products and their micro-morphology in the both different samples are examined. The microstructure of specimens is analyzed by SEM, TEM, XDR, and optical microscopy. Struvite of potassium ($MgKPO_4 \cdot 6H_2O$) is observed in the reaction products. According to the analysis, it is found that struvite exists in both crystalline and amorphous form. There is also residual magnesia in the hardened cement paste. By means of microscopy observation, it can be seen that reaction products form around the unreacted magnesia and can develop into a continuum structure, which further produces the hardened paste. Struvite can grow up to form the more perfect crystal in a long term curing age, if large enough space is available during the hydration process.

INTRODUCTION

Magnesium phosphate cements (MPCs) have excellent performance, such as rapid setting, high early strength and high adhesive properties. They are also called as chemically bonded ceramic matereials [1]. These cements have been extensively used as fast repair materials in civil engineering structures [2], [3], [4] and [5]. They have been drawing more attention in the recent years, because more and more potential applications have been realized, such as in the management of toxic waste [1] and [6], in the natural fiber composite products [7], in the fiber composites for reinforce concrete structures [8], in sealing of borehole [9], and so on.

MPCs develop its strength by means of a base-acid reaction, involving dead burned magnesia and phosphates. Usually, dead burned magnesia is a raw material for refractory, which is produced by calcination of magnesite, or abstraction from sea water. At a very high temperature, magnesium oxide becomes a highly crystallized mineral, periclase, that shows a moderate ionic potential and a relatively weak basicity. It was proved that periclase works very well with intermediate acid (usually water solution of phosphates), to form a strong binding material. In order to obtain reasonable setting time, some retarders have to be used in the base-acid reaction, usually they are compounds of boron.

In the present paper, the cementing mechanism of MPC based on the reaction of dead burned magnesia and mono-potassium phosphate is studied. It is a new type of MPC in comparison with the traditional ammonium phosphate cement. Previous studies showed that the performance of this new type of cement can be improved by fly ash, an industrial by-product from power station [10] and [11]. For example, the mortar samples containing 40% fly ash can develop to 36 MPa in compressive strength, but the mortar sample without fly ash only developed to 24 MPa in the first 4 h hydration [11].

By adding fly ash into MPC will in turn contribute to the green environment. For MPC with fly ash, the physical property, chemical property, and durability have been investigated in the previous studies [12] and [13]. The previous research also considered that the cementing action was mainly attributed to the formation of struvite of potassium. In order to further understand the cementing mechanism, the morphology of reaction products and microstructure of cement are additionally analyzed. Both fine powder and grains of dead burned magnesia are used to prepare cement paste and bonding cluster samples, respectively. The reaction products and their micro-morphology in the two different test samples are examined by X-ray diffractometer (XRD), scanning electron microscope (SEM), transmission electron microscope (TEM), and optical microscopy, respectively.

EXPERIMENT

Raw Materials

In the current study, the raw materials used are dead burned magnesia and mono potassium phosphate (MPP). The MPP is a chemical reagent in a form of a white grain crystal, which is manufactured by Guangzhou Reagent Factory (China). Two types of dead burned magnesia with the same chemical composition are used in the study. One is a fine magnesia powder (average particles size is 30.6 µm) and the other is magnesia grains (the particle size is 1 mm to 2 mm). Chemical composition of the dead burned magnesia is, MgO 89.51%, Al_2O_3 2.35%, SiO_2 4.91%, CaO 1.44% and Fe_2O_3 1.6% by weight. The chemical composition is determined by X-ray fluorescence spectrometry (XRF), and the particle size analysis is carried out by using a laser particle size distribution meter, model Counlter LS 230.

Test Methods

Preparation and Analysis of Cement Paste

The specimen of cement paste is made with fine magnesia powder, MPP and deionized water. The phosphate to magnesia ratio is 1:4, and water to dry binder ratio is 0.15. After the three ingredients were mixed, they hardened into the cement paste. Then the cement paste and its microstructure are analyzed by XRD (Philips PW1830), SEM (model JEOL-6300F), TEM (model JEM 2010), respectively.

Cementing and Analysis of Bonding Magnesia Grains

First, the saturated mono potassium phosphate (MPP) solution is prepared by solving MPP in deionize water. Second, magnesia grains are immersed into MPP solution in a small plastic bottle which is cured afterwards at a room ambient temperature. After several days, the bulk magnesia grains bond tightly together, becoming a cementing cluster. Then, the bonded magnesia cluster is taken out from the bottle, and is divided into two parts. One part is used for the study of the hydration products and microstructure of the bonded magnesia cluster. The specimen is prepared by the following procedure. The magnesia cluster is cast in epoxy in a plastic ring. After the epoxy hardened, it is polished by means of fine grinding in a progressive way. Then the specimen is coated with a carbon-palladium coating. The reaction products and microstructure are studied by optical microscopy (Olympus SZH10), XRD, SEM, respectively. To a long term observation, the other magnesia cluster part is put in a dry plastic bottle for 7 months, which allows air to go into the bottle, moisture in air can react continuously with the magnesia grains.

RESULT AND DISCUSSION

Analysis of Cement Paste

Reaction Products in the Hardened Cement Paste

The XRD pattern of hardened cement paste is shown in Fig. 1(a). According to the diffraction peaks, there are two phases of crystal products were found in the XRD pattern, one phase is unreacted periclase. The dead-burned magnesia has a high intensity, which does not react completely during hydration. It may exist as a crystalline aggregate in hardened cement paste.

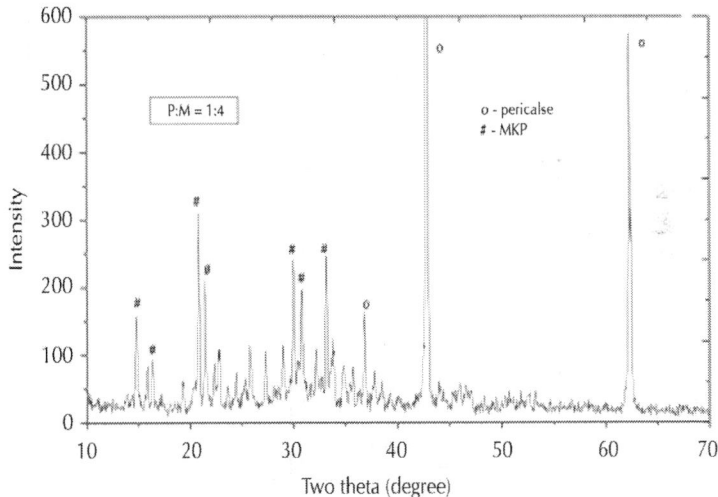

Figure 1: XRD pattern of cement paste.

The other pahse is struvite of potassium, $MgKPO_4 \cdot 6H_2O$ (MKP) which is the newly produced mineral phase inside the cement paste. According to the JCPDS system, the JCPDS card number of MKP is 35-0812, and the characteristic peaks are d=4.241 nm, 2.899 nm, 4.123 nm. At the same time, many diffused diffraction peaks exist around the main diffraction peaks of MKP. It can be inferred from the XRD analysis that the colloid species (or amorphous MKP) also appear in the hydration product. Regy and colleagues had shown that struvite exists not only as crystal form, but also exists as amorphous form, in their research work, phosphate recovery in waste water by crystallization [14].

The formation of amorphous struvite has a significant effect on the paste nature. Because the amorphous struvite have a continuous structure and can fill the voids among the grains. As a result an uninterrupted microstructure is formed. The crystal products can constitute the framework which allows amorphous mass to fill in and cohere. Like the microstructure of he hardened Portland cement, both the amorphous species (mainly calcium silicate hydrate or C–S–H gel) and crystalline (mainly calcium hydroxide and ettringite) consist jointly of the microstructure, leading to the certain mechanical properties. In MPC, the coexisting of both crystal (periclase and struvite) and amorphous species contribute to the strength development by the similar way.

SEM Analysis

Fig. 2 shows the SEM photos of cement paste after hydration for 1 h and 28 days. The microscopic photos indicate that short-bar-shape MKP products link each other, and the hardened cement paste has a tight join polycrystalline structure. The images show that the pattern of reaction products looks nearly the same after 1 h and 28 days, but the products are in a smaller size at 1 h. The growing of products may deliver a benefit effect on the strength development. Because as the hydration process continues, more reaction products are produced. Consequently a denser microstructure with a higher strength is obtained. This could explain the time-dependent mechanical properties of MPC system.

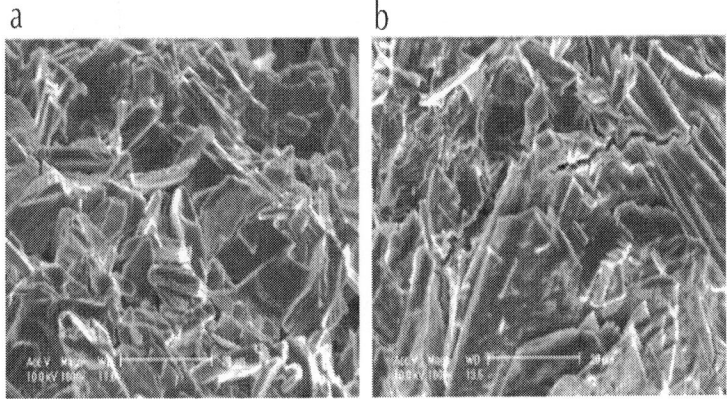

Figure 2: Microscopic analysis of hardened cement paste: (a) hardened cement paste after 1 h, (b) hardened cement paste after 28 days.

TEM Analysis

Fig. 3 is the TEM electron diffraction photos of hardened paste. The TEM electron diffraction also indicates that hydration product of the MPC is a polycrystalline composite. The polycrystalline and amorphous species have different patterns under electron diffraction in the TEM analysis. Fig. 3(a) shows the result of MPC cement paste, where the polycrystalline diffraction rings are found by TEM-SAD analysis. Polycrystalline structure consists of many crystals joining each other on the boundaries. The electron diffraction pattern of polycrystalline is concentric rings [15]. On the other hand, the amorphous substance without a regular, ordered structure of crystalline solid shows diffuse halos as far as the diffraction pattern is concerned [16]. Fig. 3(b) demonstrates the diffuse halos on the basis of the electron diffraction analysis. Therefore, the results of XRD and TEM show both crystalline and amorphous MKP exist in the hydration products.

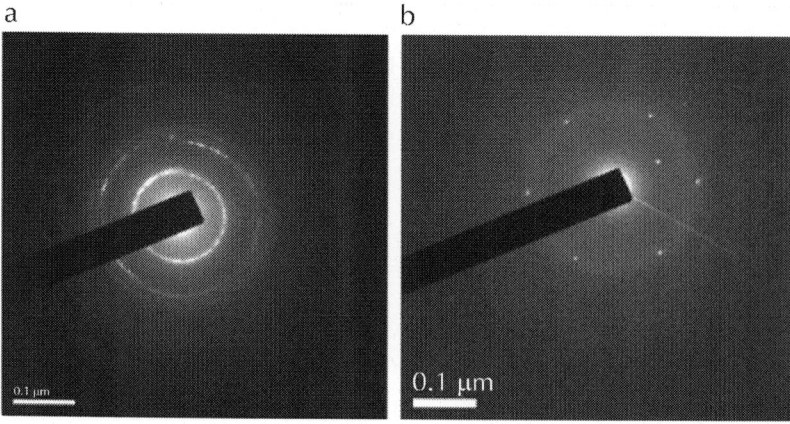

Figure 3: TEM diffraction images of sample PM4: (a) polycrystalline ring pattern with inter-planar spacings, (b) crystal diffraction points and amorphous halo.

Reaction Products among the Bonding Magnesia Grains

Observation of Binding Process

After the magnesia grains are immersed into saturated MPP solution, the reaction starts from the grain surface, more reaction products formed around magnesia grains with time.Fig. 4 illustrates the micrographs of the magnesia grains before and after they bond together. The reaction between MPP solution and magnesia in the first minutes is observed with an optical microscopy. When the saturated MPP solution is dripped on the magnesia grains, some bubbles produced immediately (see Fig. 4(a)). But the bubbles cease to appear after about 3 min.

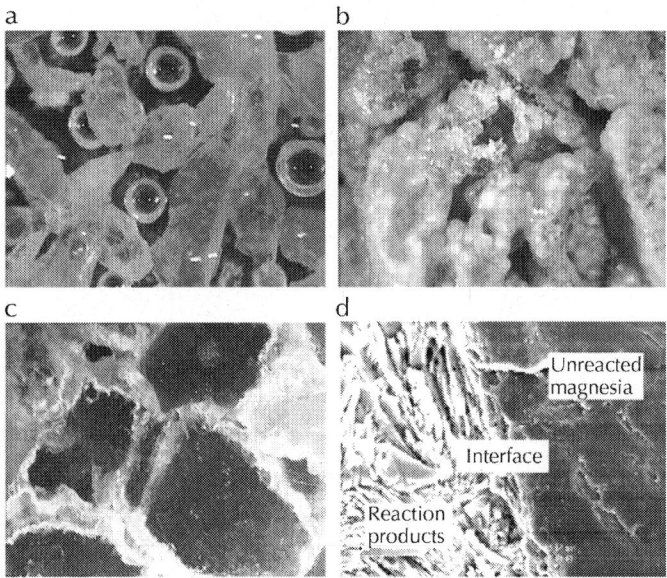

Figure 4: Binding and products formation: (a) bubbles formed when reaction (20×), (b) white products on the grains (30×), (c) products around the grains (50×), (d) interface of products and magnesia.

During immersing, magnesia is dissolved by MPP solution from surface. Struvite forms after the concentrations of magnesium, potassium (or ammonium) and phosphate ions exceed its solubility, leading to precipitation of struvite on the surface of magnesia grains. As is known, the precipitation of struvite, at a certain environment temperature, is mainly controlled by pH, supersaturation. An optimum pH for the precipitation of struvite exists, which is between 7 and 11 [15].

With immersion, hydration products accumulate more and more around magnesia grains, further bind the magnesia grains together to form a cluster. Fig. 4(b) shows that there is a layer of white reaction products developed on the surface of magnesia grains. The white reaction products are composed of fine particles. According to surface chemistry, these fine particles tend to cluster or to group together to form colloidal particles due to the reduction of the interfacial free energy (surface tension), which is attributed to the minimum surface area of sphere shape of fine particles. Observation from the direction of the bonded cross section, it is clearly demonstrated that the white products fill the voids among those magnesia grains and bond them together (see Fig. 4(c)). Fig. 4(d) shows the SEM image concerning the interface between hydration products and unreacted magnesia. In comparison with the unreacted pericalse, the newly formed structure is the amalgamation of rod-like products with voids.

According to the analysis, the binding effect is realized with the help of the white colloid orbicular products, which accumulate around the rim of the magnesia grains. These products fill the voids among grains and bind the grains together. If tiny magnesia grains with good particle size distribution are used, such as the fine magnesia powder, more reaction products will be produced, which will lead to a denser microstructure than by using large magnesia grains. Because of there are few voids in fine powder with good particle size distribution. The denser microstructure of cement paste will lead to a higher performance.

The XRD pattern of the bonded magnesia grain cluster is showed in Fig. 5. The crystalline phases in XRD patterns are unreacted magnesia and MKP. Diffusing peaks are also found here. This indicates that amorphous stuff exists. By comparing Fig. 1 withFig. 5, it is found that higher diffraction intensity of MKP presents in the bonded magnesia grain cluster. Because there is enough inter-particle space among the magnesia grains, MKP crystals can develop more perfectly. However, inside the cement paste, the MKP crystals are much finer due to the dense microstructure. Therefore, the diffraction intensity of MKP is lower in the paste sample.

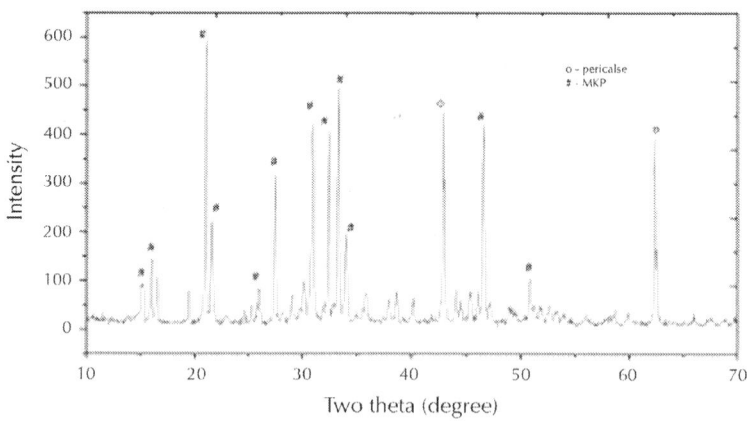

Figure 5: XRD pattern of the bonded magnesia grains.

The reaction process can be stated as the following equations.

$$KH_2PO_4 \rightarrow K^+ + H_2PO_4^-$$

$$KH_2PO_4 \rightarrow K^+ + HPO_4^{2-} + H^+$$

$$KH_2PO_4 \rightarrow K^+ + PO_4^{3-} + 2H^+$$

$$MgO + H_2O \rightarrow MgOH^+ + OH^-$$

$$MgOH^+ + 2H_2O \rightarrow Mg(OH)_2 + H_3O^+$$

$$Mg(OH)_2 \rightarrow Mg^{2+} + 2OH^-$$

$$Mg^{2+} + 6H_2O = Mg(H_2O)_6^{2+} K^+ + Mg(H_2O)_6^{2+} + PO_4^{3-} = MgKPO_4 \cdot 6H_2O$$
(MKP)

Observation of Cementing Interface of Magnesia Grain

Fig. 6(a) shows the electron image concerning the interface between hydration products and unreacted magnesia, and Fig. 6(b) illustrates the distribution of elements through the interface, which is analyzed by SEM-EDS line scanning. According the image, the rod-like products developed along the direction parallel to the surface of magnesia grains. The distribution of chemical elements in the interface shows that both element O and Mg have a similar distribution pattern. They have higher concentration in unreacted magnesia part, but lower concentration in the hydration product. Both element K and P have the similar intensity distribution in the reaction interface. The distribution intensity of Mg shows that its concentration is much lower in the hydration products than that in the unreacted magnesia part. It can be inferred that the microstructure of hydration product has a relative looser density than that of the unreacted magnesia part.

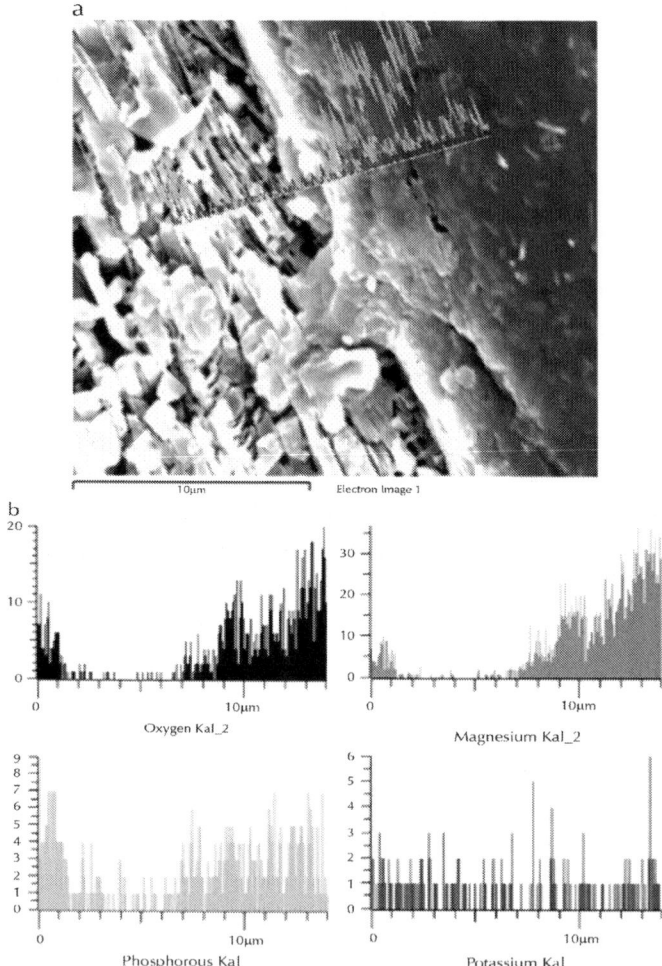

Figure 6: SEM line scanning analysis of reaction interface: (a) interface of products and magnesia, (b) distribution of elements through the interface.

Fig. 7(a) shows a magnified image of reaction products inside the bonded magnesia grain cluster. The rod-like products orient in all directions and interlock each other. According to the SEM-EDS line-scanning analysis, the element distributions of Mg, P and K demonstrate a similar intensity, which means that these three elements have the similar atomic concentration in the hydration products. This rightly matches the atomic concentration ratio of the three elements in $MgKPO_4 \cdot 6H_2O$ composition.

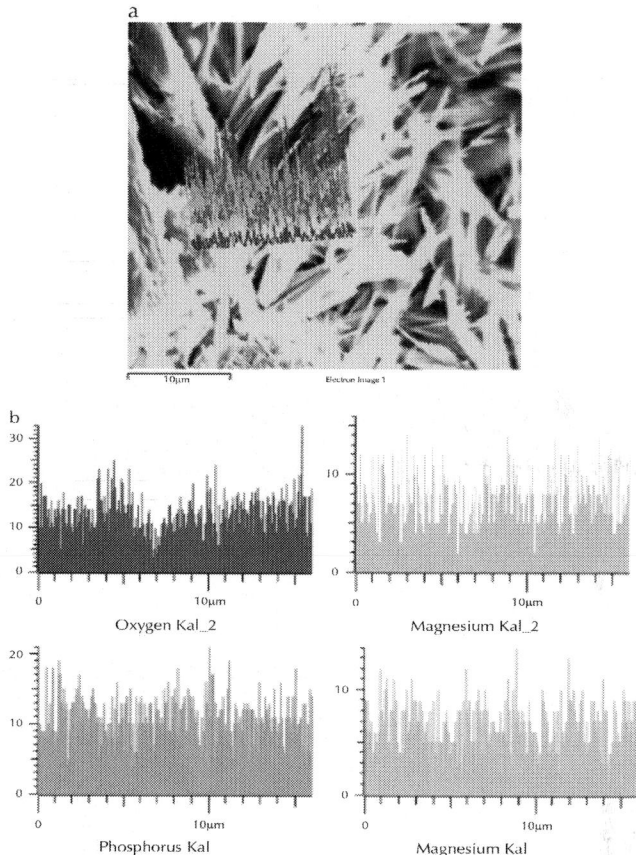

Figure 7: SEM line scanning analysis of products: (a) electron image of reaction products, (b) elements distribution in the products.

The Long Term Observation of Hydration Products

In order to observe the bonding development among the magnesia grains, one part of the boned magnesia cluster was kept in the plastic bottle. After 70 days, the specimen is observed by means of optical microscopy. It is found that the MKP crystal on the surface of magnesia grain had been grown up (see Fig. 8) into a larger size. More hydrates present around the rim of magnesia grains, and some perfect MKP crystals developed on the surface of magnesia grains.

a   b

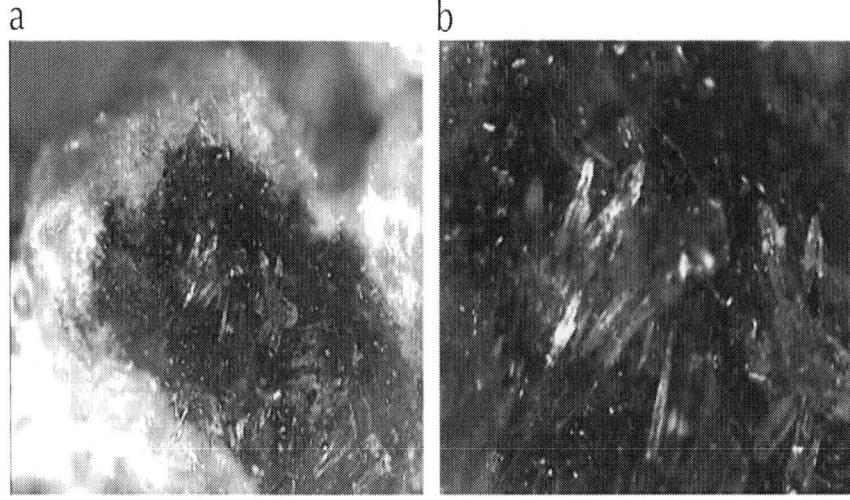

Figure 8: Products formed on the grains surface after 70 days: (a) magnification of products (25×), (b) magnification of products (70×).

CONCLUSIONS

- According to XRD, SEM and TEM analysis, the current hardened MPC paste is polycrystalline which contains struvite and unreacted magnesia. Struvite has both crystalline and amorphous forms in the hardened paste. The residual magnesia acts as fine aggregate in the hardened microstructure. The coexisting of the two forms constitutes the hardened matrix, which is the performance basis of MPC system.

- According to microscope analysis, struvite forms from the magnesia surface at first, then develops to a layer around the unreacted magnesia grain. The struvite can link and interlock each other to form a continuum structure. When the magnesia grain sizes are tiny enough, the continuum structure can develop to a denser microstructure. Denser matrix structure can be obtained by using fine magnesia powder.

- According the 70 days observation, the MKP crystals develop to large size crystal in an open space. If the condition is allowable, struvite can form perfect crystal.

ACKNOWLEDGMENTS

The authors would like to acknowledge the financial support provided by National Natural Science Funds (51172146), National Natural Science Funds for Distinguished Young Scholar (50925829) and Natural Science Foundation for the team project of Guangdong Province (No. 9351806001000001).

REFERENCES

1. D.M. Roy, New strong cement materials: chemically bonded ceramics, Science 235 (1987) 651–658.

2. S.S. Seehra, S. Gupta, S. Kumar, Rapid setting magnesium phosphate cement for quick repair of concrete pavements— Fig. 8. Products formed on the grains surface after 70 days: (a) magnification of products (25), (b) magnification of products (70). Z. Ding et al. / Ceramics International 38 (2012) 6281– 6288 6287 characterization and durability aspects, Cement and Concrete Research 23 (1993) 254–266.

3. A.K. Sarker, Phosphate cement-based fast-setting binders, Ceramic Bulletin 69 (1990) 234–238.

4. S. Popovics, N. Rajendran, M. Penko, Rapid hardening cements for repair of concrete, ACI Materials Journal 84 (1987) 64–73.

5. Q.B. Yang, B.R Zhu, S.Q. Zhang, X.L. Wu, Properties and applications of magnesia-phosphate cement mortar for raped repair of concrete, Cement and Concrete Research 30 (2000) 1807–1813.

6. A.S. Wagh, R. Strain, S. Jeong, D. Reed, T. Krouse, D. Singh, Stabilization of rocky flats Pu-contaminated ash within chemically bonded phosphate ceramics, Journal of Nuclear Materials 265 (1999) 295–307.

7. K. Patrick, Donahue, D.A. Matthew, Durable phosphate-bonded natural fiber composite products, Cons and Bulletin of Materials 24 (2010) 215–219.

8. Z. Ding, Z.X. Lu, Y. Li., Feasibility of Basalt fiber reinforced inorganic adhesive for concrete strengthening, Advances in Materials Research 287–290 (2011) 1197–1220.

9.   Singh, D., Perry, A.S. and Jeong, S. Y., Pumbale/Injectable Phosphate-bonded Cements, US Patent. US6, 204,214. Mar. 20. 2001.

10. Wagh A.S., Jeong, S.Y., D., Singh, High Strength Phosphate Cement using Industrial Byproduct Ashes, in: Azizinannini, A., et al. (Ed.), Proceedings of First international Conference, pub. Amer. Soc. Civil Eng. 1997, pp. 542–533.

11. Z. Ding, Z.J. Li, High-early-strength magnesium phosphate cement with fly ash, ACI Materials Journal 102 (2005) 375–381.

12. Z. Ding, Z.J. Li, Effect of aggregates and water contents on the properties of magnesium phosphor silicate cement, Cement and Concrete Composites 27 (2005) 11–18.

13. Z. Ding, Z.J Li, Xing Feng, Chemical durability investigation of magnesium phosphosilicate cement, Key Engineering Materials 302–303 (2006) 275–281.

14. Regy, S., Mangin, D., Klein, J.P. and Lieto, J., Phosphate Recovery in Waste Water by Crystallization. /http://www.nhm.ac.uk/mineral ogy/phos/LagepReport.PDFS. 2002.

15. N.R. Yang, Characterization of Inorganic Materials, Wuhan. Press of Wuhan University of Industry, 1990 141.

16. Y. Zhang, K. Hono, A. Inoue, A. Makino., T. Sakurai, Nanocrystalline structure evolution in Fe, Zr, B, soft magnetic materials, Acta Materialia 44 (1996) 1497–1510.

6

Ceramic Waste as Aggregate and Supplementary Cementing Material: A Combined Action to Contrast Alkali Silica Reaction (ASR)

Maria Chiara Bignozzi and Andrea Saccani

Dipartimento di Ingegneria Civile, Ambientale e dei Materiali, Via Terracini 28, 40131 Bologna, Italy

ABSTRACT

Recently, many efforts have been made to recycle waste of different nature as constituents of sustainable concrete. This practice produces large environmental benefits that can be further extended if deleterious chemical side-reactions, deriving from the use of some types of waste and/or raw materials, could be prevented and suppressed. This paper presents the combined action of different ceramic wastes partially replacing natural sand and cement, respectively. Alkali silica reaction (ASR) promoted by boron–silicate and lead–silicate glass used as fine aggregates (≤ 4 mm) is limited and controlled by using a new type of blended cement based on a siliceous residue coming from sludge produced by the polishing of porcelain stoneware tiles. The results of expansion tests carried out in accelerated conditions together with mechanical and microstructure characterisations of mortar samples highlight the combined action of the investigated wastes. Indeed, the blended cement containing porcelain stoneware polishing residue can be effectively exploited as valid alternative to pozzolan cement.

INTRODUCTION

Construction materials, concrete in particular, have attracted great interest as a possible way to recycle industrial and/or urban waste. The relatively simple concrete mixing technology, the large volumes of produced material, the uniform and widespread diffusion of building sites on the territory can easily furnish accessible sites without drawbacks deriving from long transport routes. Fine/coarse aggregates and cement can be successfully replaced by different kind of waste. Recycled aggregates coming from construction and demolition waste have been investigated [1], [2], [3] and [4] as well as scrap tyres [5] and [6], plastic waste of different origin [7], [8] and [9] and waste glass [10] as fine aggregates. As a partial substitution of ordinary Portland cement (OPC), carbon fly-ash has been the precursor of a new type of supplementary cementing materials (SCMs), followed by the use of blast furnace slag [11], pulverized soda lime glass [12], residuals of glass separate collection [13], treated bottom ashes [14] and biomass fly ash [15], activated slag [16] and [17], rice husk ash [18], metakaolin and calcined clays [19] and even ceramic residues [20]. In the above reported studies waste is proposed as secondary raw material, thus avoiding the costs associated to waste disposal and the negative impact on the environment (landfill saturation, land pollution, etc.). However, the use of non-traditional raw materials can lead to unexpected disadvantages as undesired chemical reactions can occur between traditional and recycled constituents. Indeed, recycled SCM can interfere with cement hydration reactions and mechanical strength development. For instance, metallurgical slags (blast furnace slag, steel slag, etc.) partially replacing OPC, usually result to blended cements with lower early strength and longer setting time [21] and [22]. Municipal solid waste incineration (MSWI) fly ash has some cementitious activity, but the reactivity is relatively low and its addition to cement may lead to retardation of cement hydration [23]. Special treatments are usually carried out on MSWI fly ash to make them suitable as SCM [24] and [25]. Furthermore, recycled aggregates can promote the formation of delayed ettringite and/or alkali silica reaction (ASR) leading to the growth of expansive products with disruptive consequences on concrete elements [26], [27] and [28].

ASR is a deleterious reaction between the amorphous silica of aggregates and cement alkalis, leading to expansive gel formation. Several studies are addressed to better understand ASR when glass is used as sand or gravel replacement in concrete production. Indeed, it has been recently demonstrated that the expansion extent strongly depends on glass chemical composition [26]. For example, lead–silicate and boron–silicate rich glass when used as sand replacement (25 wt. %) in mortar samples induce ASR. The elevate content of ions such as Pb^{2+} and B^{3+} lead to a high solubility of the relevant glass in cement environment and the development of highly expansive products. As recycled glass used for natural aggregate replacement may derive from many sources (urban separated collection, industrial waste, pharmaceutical glassware, television screen, fluorescent lamps, etc.) the possibility that glass rich in Pb or B ions might be used in concrete production is rather high, unless a very accurate selection of cullet is not carried out.

One of the most frequently applied methods to inhibit ASR is to use pozzolan cement (CEM IV) or blended cement with pozzolan supplementary cementing materials [29], [30], [31] and [32]. However, these types of cements are not commonly used as they are generally more expensive than OPC. Certainly, CEM IV pozzolan cement represents only about 6% of the total European cement production [33].

Recently [20] and [34], a solid residue (hereafter named as PR, polishing residue) resulting from porcelain stoneware tiles polishing sludge has been investigated as a new type of SCM. PR is the solid part of the relevant sludge formed during porcelain stoneware polishing step. This operation is performed on fired tiles by abrasive devices made of silicon carbide (SiC) and magnesium-based (MgOHCl) binder. In Europe, polishing sludge is classified as non-hazardous waste (European Waste Code 10.12.99), but the presence of some compounds (CaO, MgO, SiC and chlorine compounds coming from the abrasive tools) prevents sludge re-introduction into the ceramic production cycle compromising a closed loop recycling. Italy, the 3rd world tile producer, disposes to landfill more than 20,000 ton of polishing sludge every year [35]. Ceramic materials production is largely diffuse in Europe where the major producing countries are Spain, Italy, Germany, UK and France. Between the new Member States of the EU, production is well developed in the Czech Republic, Poland and Hungary, which all have strong ceramics sector [36]. Indeed, ceramic industry also generates large amounts of waste that can be fruitfully exploited for the production of sustainable building materials [37] and [38].

Polishing residue is obtained by drying the relevant sludge and eliminating the fraction greater than 0.106 mm, that is usually less than 5 wt.%. Neither crushing milling treatments nor washing procedures are necessary, thus ensuring a low cost of the residue as an alternative cement constituent.

The new binder made up of 25 wt. % PR + 75 wt. % CEM I 52.5 R exhibits mechanical and durability performances comparable/better than those shown by mortar based on OPC (100% CEM I 52.5R) and a chemical activity of PR towards calcium hydroxide was determined [20]. The aim of this work is to evaluate the effect of PR based binder as an ASR inhibitor when recycled glass rich of Pb or B ions are used as 25 wt. % replacement of natural sand, thus exploiting different ceramic wastes largely produced Promoting PR as SCM able to mitigate the deleterious effects of ASR means encouraging the use of blended cement with low clinker content and, consequently, moving towards the use of sustainable construction materials.

The behaviour of PR based binder has been compared with that of OPC (CEM I) and pozzolan cement (CEM IV/A). With each types of cement, mortar samples were prepared using natural sand and lead–silicate or boron–silicate glass as fine aggregate. The results of ASR test, evaluated in accelerated condition (ASTM C1260 [39]), as well as mechanical properties and microstructure of mortar samples cured in standard conditions are reported and discussed disclosing the role played by the investigated wastes.

EXPERIMENTAL PART

Materials

A porcelain stoneware tiles polishing sludge was used to obtain PR residue. A highly homogenous sample of sludge representative of a single collecting site was selected (Hera S.p.A. Service for the Environment, Modena, Italy). The sludge was dried (T = 105 °C for 24–36 h) and ground in a laboratory ball mill to obtain a fine grain size distribution (D_{10} = 1.5 µm, D_{50} = 8 µm and D_{90} = 30 µm), as characterised by laser particle-size analyser (Master Sizer 2000, Malvern Instrument). As this residue was investigated as cement constituent, the fraction >106 µm was eliminated by sieving. The resulting material, henceforth named PR, has a specific surface area of 25.9 m²/g (by BET analysis Micrometrics Gemini Series, 2360 with N_2 at 77°K). By use of an X-ray fluorescence spectrometer (PW1414, Philips) and X-ray powder diffraction (PW3710, Philips), the chemical and mineralogical composition of PR was determined. PR is mainly constituted by SiO_2 = 62.2; Al_2O_3 = 15.8, MgO = 6.8, Na_2O = 3.7, CaO = 2.2, K_2O = 1.5 and ZrO_2 = 1.2 wt. %. Fig. 1 schematically shows where PR is located in the $CaO–SiO_2–Al_2O_3$ system in comparison with the most used SCM.

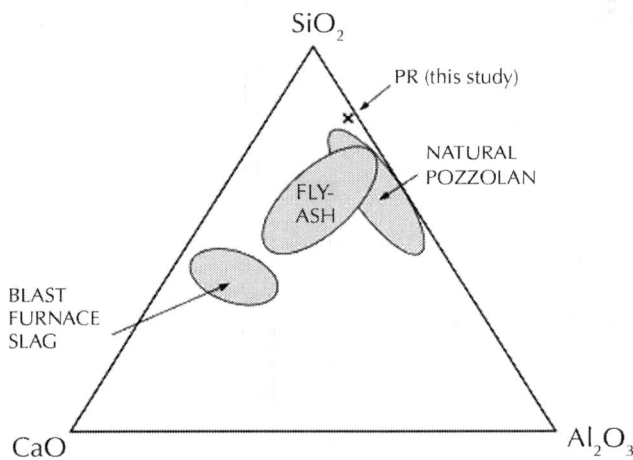

Figure 1: Ternary diagram of mostly used SCM.

By mineralogical analysis, PR discloses the presence of quartz, mullite and albite calcian, deriving from porcelain stoneware body. Amorphous SiO_2 phase, traces of calcite and silicon carbide, deriving from the abrasive tools, has also been determined. A complete chemical–physical characterisation of PR is reported elsewhere [20].

PR based binder was prepared by dry mixing 25 wt.% PR and 75 wt.% of CEM I 52.5 R: the amount of PR was set on the basis of CEM II blended cement compositional range reported in EN 197-1 [40]. Mixing procedure was carried out in a mixer for 30 min to ensure a repeatable mix. Chemical (chloride and sulphate content) and physical (initial setting time and soundness) parameters of PR based binder were previously determined [20] to ascertain the fulfilling of cement requirements set by EN 197-1.

Reference binders were 100% OPC CEM I 52.5 R and 100% pozzolan cement CEM IV/A 42.5 R (Italcementi, Italy)

Silica sand, with normalized particle size distribution according to EN 196-1 [41], was used as aggregate for mortar preparation. Three different types of glass waste were used as a partial substitution (25 wt. %) of silica normalized sand: lead–silicate glass (CG), coming from production of tableware, giftware and home décor items in crystal (kindly supplied by CALP, Colle di Val d'Elsa (Italy), boron–silicate glass (BG) and amber boron–silicate glass (ABG) coming from pharmaceutical containers (kindly supplied by Bormioli Rocco, Fidenza (Italy). The chemical composition (wt. %) of glass is reported in Table 1: values below 50 × LLD have been omitted. The investigated glass was dry ground in a laboratory steel jaw crusher to get particles between 0.075 and 2.00 mm, with size distribution close to that of normalized sand (EN 196-1). Fineness modulus, determined according to EN 12620 [42], is 2.7 for BG, 2.6 for ABG, 2.8 for CG and 2.8 for normalized sand, respectively.

Table 1: Chemical analysis (wt. %) of the investigated glass

	BG[a]	ABG[b]	CG[c]
SiO_2	68.12	65.30	61.70
Al_2O_3	5.64	5.88	0.04
TiO_2	–	2.89	–
Fe_2O_3	0.02	0.78	–

CaO	1.53	1.42	0.01
MgO	–	–	0.03
K_2O	1.24	1.16	–
Na_2O	8.06	8.02	3.96
ZrO_2	–	–	–
NiO	–	–	–
ZnO	0.82	0.78	0.90
BaO	2.95	2.86	–
PbO	–	–	25.30
P_2O_5	–	–	–
B_2O_3	11.60	10.70	0.72

a BG: boron–silicate glass.

b ABG: amber boron–silicate glass.

c CG: lead–silicate glass.

Mortar Preparation

The investigated mix designs and relevant acronyms are reported in Table 2: mortar samples containing 100 wt.% of silica sand or 75 wt.% of silica sand +25 wt.% of glass, were alternatively prepared with 100% CEM I, 100% CEM IV/A and 25% PR + 75% CEM I 52.5. Mortar samples have been named M, followed by CEM I, CEM IV or PR according to the binder used in the mix design and, when necessary, by BG, ABG and CG indicating the glass type replacing 25 wt.% of silica sand (e.g. MPR-BG is the mortar prepared with PR based binder and boron–silicate glass replacing 25% of silica sand).

Table 2: Mix-design of mortar samples

	Binder			Aggregate			
	CEM I	CEM IV	25%PR + 75%CEM I	Silica sand	BG	ABG	CG
MCEM I	450.0	–	–	1350.0	–	–	–
MCEM I-BG	450.0	–	–	1012.5	337.5	–	–

MCEM I-ABG	450.0	–	–	1012.5	–	337.5	–
MCEM I-CG	450.0	–	–	1012.5	–	–	337.5
MCEM IV	–	450.0	–	1350.0	–	–	–
MCEM IV-BG	–	450.0	–	1012.5	337.5	–	–
MCEM IV-ABG	–	450.0	–	1012.5	–	337.5	–
MCEM IV-CG	–	450.0	–	1012.5	–	–	337.5
MPR	–	–	450.0	1350.0	–	–	–
MPR-BG	–	–	450.0	1012.5	337.5	–	–
MPR-ABG	–	–	450.0	1012.5	–	337.5	–
MPR-CG	–	–	450.0	1012.5	–	–	337.5

Mortar samples were prepared by means of a Hobart planetary mixer of about 5 L capacity, according to the normalized mix-design (binder, sand and water in a weight ratio of 1:3:0.5) and procedure for cement mechanical strength determination (EN 196-1). Mortar prisms (40 × 40 × 160 mm) for mechanical strength determination were cured at 20 °C and R.H. > 95% for 28 days. Mortar prisms (40 × 40 × 160 mm) for accelerated expansion tests were cured in NaOH 1 M at 80 °C for 14 days, according to ASTM C1260.

Characterisation

Mortar workability was determined by using a flow table according to EN 1015-3 [43]. Compressive strength of mortar samples was determined by an Amsler–Wolpert machine (100 kN) at a constant displacement rate of 2400 ± 200 N/s. Mechanical data are reported as average of at least five measurements. Expansion measurements were determined by means of a mechanical comparator (0.001 mm accuracy) on samples cooled at room temperature: the detailed procedure is elsewhere reported [26] and close to that prescribed by ASTM C1260. Pore size distribution measurements were carried out by mercury intrusion porosimeter (MIP, Carlo Erba 2000) equipped with a macropore unit (Model 120, Fison Instruments). Porosimeter samples, about 1 cm^3, were cut by a diamond saw, dried under vacuum and kept under P_2O_5 in a vacuum dry box till testing. Scanning electron microscopy (SEM) analysis was carried out by means of XL 20 Philips instrument. Sample fresh fractured surfaces were coated by graphite. Phase recognition was acquired by energy-dispersive X-ray spectroscopy (EDX Genesis 2000).

RESULTS AND DISCUSSION

Fresh State Behaviour

The use of the investigated waste has a direct effect on mortar workability as reported in Fig. 2. Regardless the aggregate type and comparing only the effect of the binder, CEM IV based samples always exhibit the highest values of workability, whereas PR based samples always lead to the lowest results. Such behaviour can be ascribed to the different shape of binder particles (Fig. 3). PR particles, deriving from mechanical abrasion processes, mainly have angular shape thus hindering particles flowing. On the contrary, natural pozzolan of volcanic origin and/or silica fly ash, usually both present in CEM IV, mainly exhibit rounded contours particles, which favour workability.

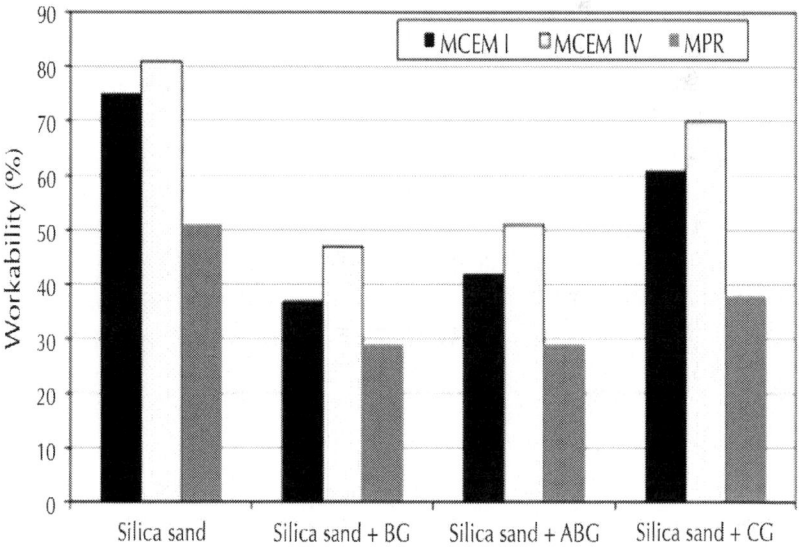

Figure 2: Workability of mortar samples.

(a)   (b)

Figure 3: SEM images of CEM IV (a) and PR (b).

Glass introduction as 25% sand replacement further contributes to a general decrease in workability. The workability loss is more evident when BG and ABG are used according to their lower value of fineness modulus. Moreover, all the glass types have angular shapes with very sharp edges due to their brittle nature, thus explaining the results detected for the relevant mortar samples.

Expansion, Mechanical and Microstructure Results after Accelerated Curing Condition

With the aim to prove if a combined action occurs between glass waste and PR based binder, expansion test has been carried out in accelerated conditions on mortar specimens alternatively prepared with BG, ABG, CG and normalized sand as fine aggregate and CEM I, CEM IV and 75% CEM I + 25% PR as binder.Fig. 4 report the expansion data after 14 days of curing in NaOH 1 M solution at 80 °C.

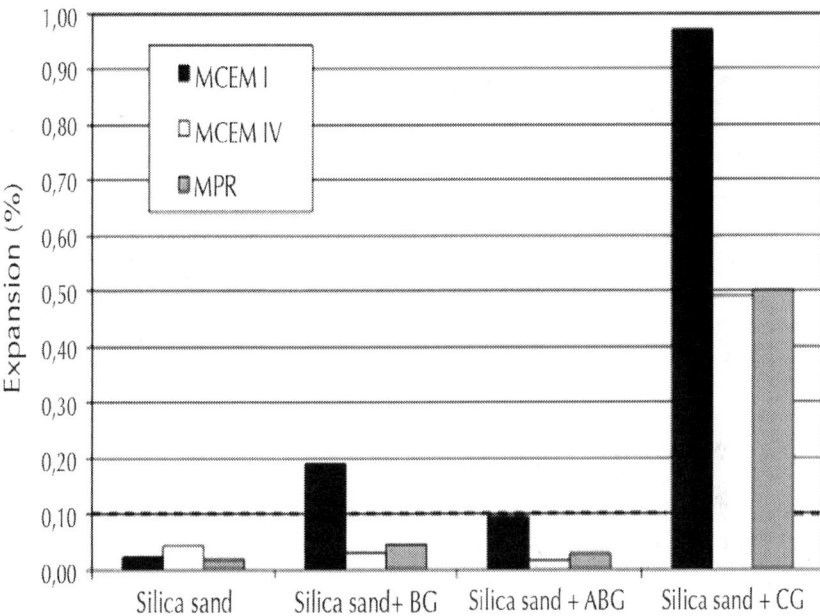

Figure 4: Expansion of mortar samples after 14 days of curing in NaOH 1 M solution at 80 °C (dotted line shows the limit set up by ASTM C1260; expansion values reported are the average of three measurements).

According to previous results [26], when CEM I is used with BG, ABG and CG, expansion processes occur in the relevant mortar samples generally overcoming the limit (0.10%) fixed by ASTM C1260 (MCEMI-ABG shows expansion equal to 0.10%). CG causes the largest mortar expansion. When CEM IV and PR based binder are adopted, a strong reduction of expansion process is evident. Mortar samples containing BG and ABG are safely well below the limit and an expansion decrease of about 50% is determined when CG is used. Thus, PR based binder works as inhibitor of alkali silica reaction as commercial pozzolan cement does.

Fig. 5 shows compressive strength values of the investigated mortars after expansion test.

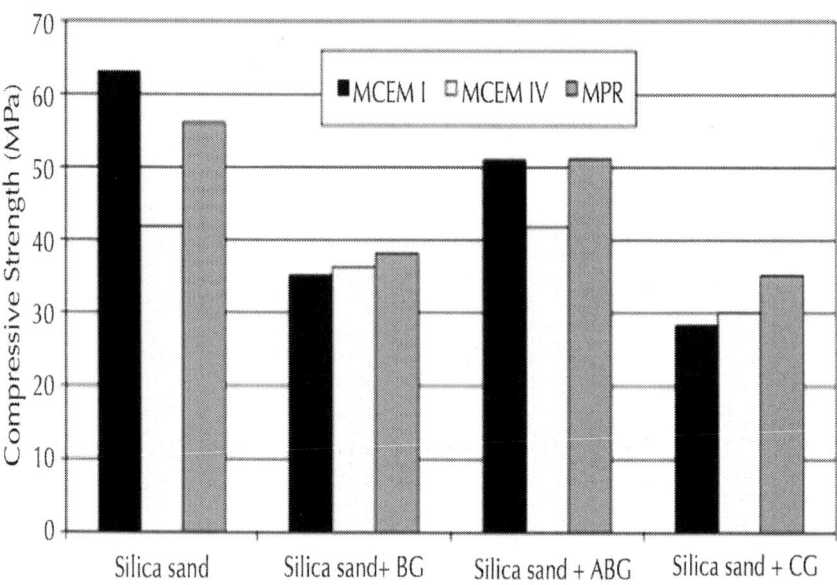

Figure 5: Compressive strength of mortar samples after expansion test.

According to the results above discussed, the highest compressive strengths are determined when no expansion occurs. When 100% silica sand is used as aggregate, the mortar compressive strengths are ≥ 52.5 MPa and ≥ 42.5 MPa in good agreement with the mechanical strength class of the cement used (CEM I 52.5 R, CEM IV 42.5 R, to 75% CEM + 25% PR a 52.5 R class has been previously assigned [20]). When glass is used as 25% sand replacement, the decrease in compressive strength is strictly related to the extent of the expansion. For MCEM I, the highest reduction is detected when CG is adopted, followed by samples containing BG and ABG. For MCEM IV and MPR the compressive strength reduction is also more evident when CG is used according to expansion results that still overcome the limit of 0.10%. However, for MCEM IV and MPR containing BG and ABG the lowering of the compressive strength is less evident according with the very low expansion determined.

SEM analysis provides an insight in mortar microstructure after the expansion test. Fig. 6 shows BG and CG in the three different cement matrices. The sharp, un-reacted contour of BG and the good adhesion with PR based matrix confirms the suppression of ASR (Fig. 6a). A similar behaviour for BG is also observed when it is immersed in CEM

IV matrix (Fig. 6b), whereas BG reactivity is clearly evident when CEM I is used (Fig. 6c). The evidence of BG reacted particles and cracks in OPC matrix are further proofs of ASR occurring in MCEM I sample.

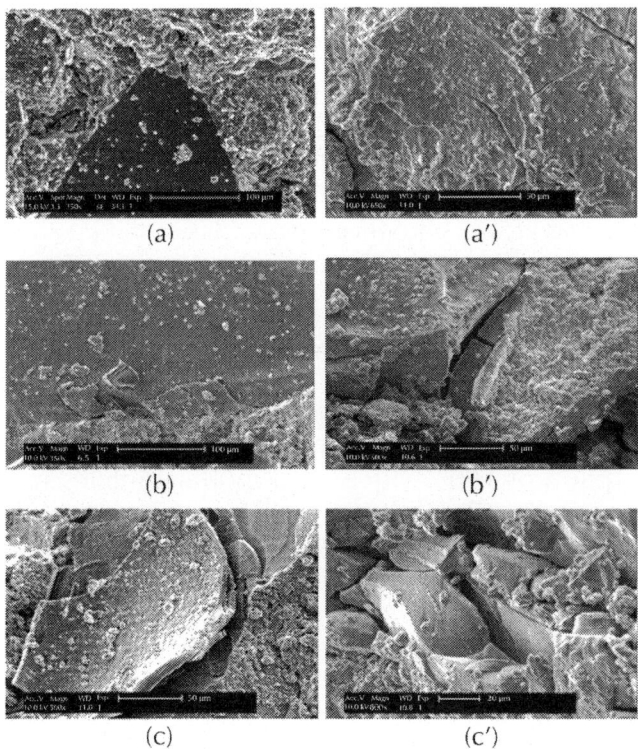

Figure 6: Microstructure of MPR-BG (a) and MPR-CG (a'), MCEM IV-BG (b) and MCEM IV-CG (b'), MCEM I-BG (c) and MCEM I-CG (c') after expansion test.

The reactivity of CG, which always forms expansive products, becomes more and more evident passing from MPR (Fig. 6a) and MCEM IV (Fig. 6b) to MCEM I matrix (Fig. 6c). For the latter, the extended presence of expansive products on CG surface leads to the formation of very large cracks thus exhibiting the typical effects of ASR. Indeed, PR based binder is able to reduce ASR caused by BG and ABG under the limit and mitigate ASR consequences when CG is used. PR based binder behaviour is then close to that of commercial pozzolan cement and, consequently, PR use as a SCM is promising.

Mechanical and Microstructure Results after Standard Curing Condition

The investigated types of waste are attractive as their use can be exploited for green/sustainable building. For this reason mechanical characterisation and microstructure analyses have been carried out on mortar samples cured in standard conditions too.

Compressive strengths are reported after 28 days of curing at 25 °C and R.H. ≥ 90% in Fig. 7. Compressive strength of mortar samples strictly follows binder strength class. In fact, MCEM I and MPR (strength class: 52.5 R) show the highest values comparing to MCEM IV (strength class: 42.5 R). Glass introduction, regardless the type, leads to a compressive strength lowering and such reduction is particularly evident when BG is used. Compressive strength decrease is probably due to the low workability detected when glass is used as aggregate which leads to a general increase in total open porosity (water absorption of 100% silica sand mortar samples is about 10% lower than relevant mortars containing glass). Moreover, less effective adhesion of the binder matrix to the smooth surface of the glass has been elsewhere reported [44] and [45].

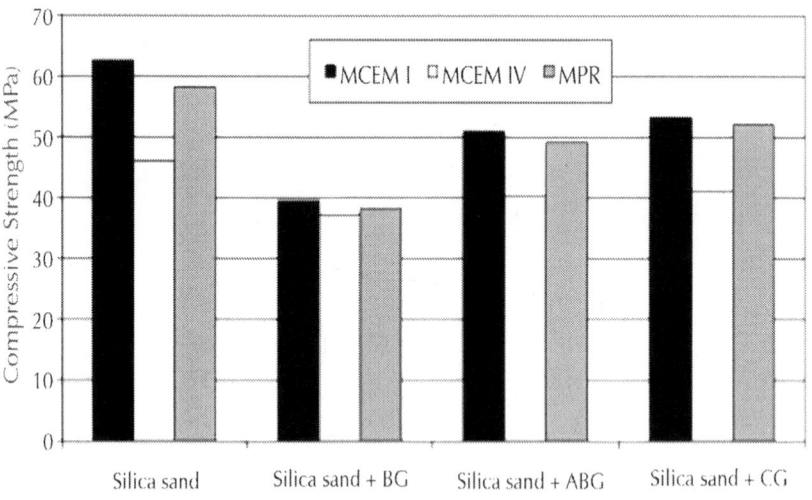

Figure 7: Compressive strength of mortar samples after 28 days of curing in standard condition.

Compressive strengths data obtained with standard curing condition are generally equal or superior than the relevant results obtained in accelerated curing condition. The different behaviour in mechanical performances is particularly evident when CG is used as aggregate, according to CG ability to promote ASR.

Fig. 8 and Fig. 9 report pore size distributions obtained by MIP for MPR and MCEM IV samples, respectively. Regardless the binder used, mortar samples containing BG exhibit the highest porosity for pore size ≥0.1 μm according to the lowest workability detected (Fig. 2). Cumulative pore volume data of MPR samples are lower than those of relevant samples prepared with CEM IV. The mean pore radius for MPR ranges from 0.05 to 0.07 μm, whereas for MCEM IV values between 0.09 and 0.13 μm are detected. The more refined capillary porosity suggests a faster hydration at 28 day of curing for PR based binder compared to the commercial pozzolan binder. Indeed, PR works as a very active SCM promoting a more compact matrix in standard curing condition and inhibiting/moderating ASR.

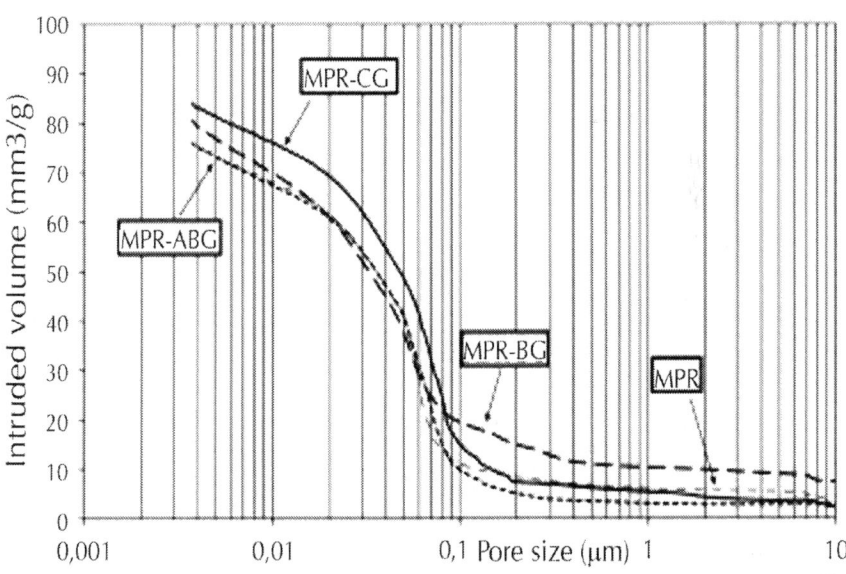

Figure 8: Pore size distribution of MPR samples after 28 days of curing in standard condition.

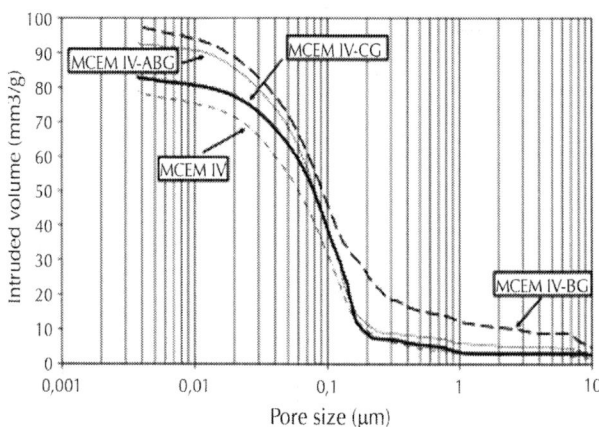

Figure 9: Pore size distribution of MCEM IV samples after 28 days of curing in standard condition.

Fig. 10 shows microstructure and EDX analysis of the fracture surfaces of MCEM IV and MPR samples after 28 days of curing. No significant differences in matrixes morphology can be highlighted. EDX results on hydration products show an average Ca/Si weight ratio lower than 1.5, in particular 1.3–1.4 for MCEM IV and 1.1–1.3 for MPR. The determined ratios are comparable and confirm that PR particles react in a very similar way to natural pozzolan, even though low amounts (<3%) of Mg, K and Na are present in MPR.

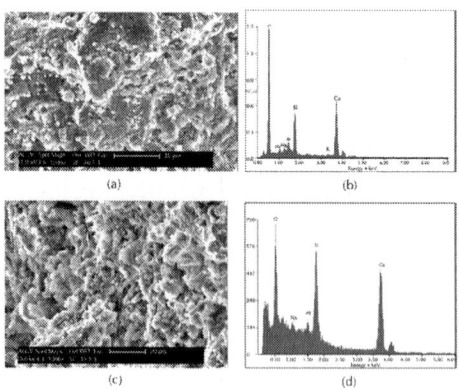

Figure 10: Microstructure and EDX analysis of PR matrix (a and b) and CEM IV matrix (c and d) after 28 days of curing in standard condition.

CONCLUSIONS

Based on the results above described, the following conclusions can be drawn by the present study:

- Polishing residue (PR) coming from porcelain stoneware tiles polishing sludge is effective in reducing/suppressing alkali silica reaction promoted by the use of glass aggregate;
- Mortar samples prepared with PR based binder and boron or amber-boron silicate glass show good dimensional stability, high mechanical properties and compact microstructure thus ensuring their use for sustainable building;
- PR based binder is a valid alternative to commercial pozzolan cement as it is able to sustain a safe use of cullet coming from many sources (separated collection, industrial waste, fluorescent lamps, etc.) potentially containing fractions of expanding glass.

It has been highlighted that a tailored combination of different types of ceramic waste can be successful for designing green building materials. As raw materials safeguard and waste exploitation create environmental benefits and represent the milestones for a sustainable development, political authorities should endorse such issues with the aim to promote innovative and ecological attitudes in the construction industry.

Finally, this research promotes the use of supplementary cementing material for ASR suppression thus limiting such deleterious reaction that generally creates severe and dangerous damages in the building construction. Potentially active aggregates as glass or amorphous silica rich natural minerals could thus be safely used for mortar and concrete preparation.

REFERENCES

1.   Marinkovic S, Radojanin V, Maselev N, Ignjatovic I. Comparative environmental assessment of natural and recycled aggregate concrete. Waste Manage 2010; 30:2255–64.

2.   Lopez-Gayarre F, Serna P, Domingo-Cabo A, Serrano Lopez MA, Lopez Collina C. Influence of recycled aggregate quality and

proportioning criteria on recycled concrete properties. Waste Manage 2009; 29:3022–8.

3. Sani D, Moriconi G, Fava G, Corinaldesi V. Leaching and mechanical behaviour of concrete manufactured with recycled aggregates. Waste Manage 2005; 25:177–82.

4. Topcu IB. Physical and mechanical properties of concretes produced with waste concrete Cem Concr Res 1997; 27:1817–23.

5. Aiello MA, Leuzzi F. Waste tyre rubberized concrete: properties at fresh and hardened state. Waste Manage 2010; 30:1696–704.

6. Bignozzi MC, Sandrolini F. Tyre rubber waste recycling in self-compacting concrete. Cem Concr Res 2006; 36 (4):735–9.

7. Choi YW, Moon DJ, Kim YJ, Lachemi M. Characteristics of mortar and concretes containing fine aggregates manufactured from recycled PET bottles. Constr Build Mater 2009; 23 (8):229–35.

8. Moncef N, Sumner J. Recycling waste latex paint in concrete. Cem Concr Res 2003; 33:857–63.

9. Siddique R, Khatib J, Kaur I. Use of recycled plastic in concrete: a review. Waste Manage 2008; 28 (10):1835–52.

10. Limbachiya MC. Bulk engineering and durability properties of washed sand concrete Constr Build Mater 2009; 23:1078–83.

11. Menéndez G, Bonavetti V, Irassar EF Strength development of ternary blended cement with limestone filler and blast-furnace slag. Cem Concr Compos 2003; 25:61–7.

12. Shi C, Wu Y, Riefler C, Wang H. Characteristics and pozzolanic reactivity of glass powders. Cem Concr Res 2005; 35:987–93.

13. Bignozzi MC, Saccani A, Sandrolini F. Matt waste from glass separated collection: an eco-sustainable addition for new building materials. Waste Manage 2009; 29:329–34.

14. Saccani A, Sandrolini F, Andreola F, Barbieri L, Corradi A, Lancellotti I. Influence of the pozzolanic fraction obtained from vitrified bottom-ashes from MSWI on the properties of cementitious composites. Mater Struct 2005; 38:367–71.

15. Wang S, Baxter S, Fonseca F. Biomass fly ash in concrete: SEM, EDX and ESEM analysis. Fuel 2008; 87:372–9.

16. Frías M, Rodríguez C. Effect of incorporating ferroalloy industry wastes as complementary cementing materials on the properties

of blended cement matrices. Cem Concr Compos 2008; 30:212–9.

17. Li C, Sun H, Li L. A review: comparison between alkali-activated slag (Si + Ca) and metakaolin (Si + Al) cement. Cem Concr Res 2010; 40:1341–9.

18. Ganesan K, Rajagopal K, Thangavel K. Rice husk ash blended cement: assessment of optimal level of replacement for strength and permeability properties of concrete. Constr Build Mater 2008; 22:1675–83.

19. Sabir BB, Wild S, Bai J. Metakaolin and calcined clays as pozzolan for concrete: a review. Cem Concr Compos 2001; 23:441–54.

20. Andreola F, Barbieri L, Lancellotti I, Bignozzi MC, Sandrolini F. New blended cement from polishing and glazing ceramic sludge Inter J Appl Ceram Technol 2010;7 (4):546–55.

21. Shi C, Qian J. High performance cementing materials from industrial slags – a review. Resour Conserv Recy 2000; 29:195–207.

22. Shi C, Day RL. Early strength development and hydration of alkali-activated blast furnace slag/fly ash blends Adv Cem Res 1999; 11:189–96.

23. Shi H, Kan L. Characteristics of municipal solid wastes incineration (MSWI) fly ash–cement matrices and effect of mineral admixtures on composite system. Constr Build Mater 2009; 23:2160–6.

24. Lee TC, Wang WJ, Shih PY, Lin KL. Enhancement in early strengths of slag cement mortars by adjusting basicity of the slag prepared from fly-ash of MSWI. Cem Concr Res 2009; 39:651–8.

25. Gao X, Wang W, Ye T, Wang F, Lan Y. Utilization of washed MSWI fly ash as partial cement substitute with the addition of dithiocarbamic chelate. J Environ Manage 2008; 88:293–9.

26. Saccani A, Bignozzi MC. ASR expansion behavior of recycled glass fine aggregates in concrete. Cem Concr Res 2010; 40:531–6.

27. Saikia N, Cornelis G, Mertens G, Elsen J, Van Balen K, Van Gerven T, et al. Assessment of Pb-slag, MSWI bottom ash and boiler and fly ash for using as a fine aggregate in cement mortar. J Hazard Mater 2008; 154:766–77.

28. Shehata MH, Christidis C, Mikhaiel W, Rogers C, Lachemi M. Reactivity of reclaimed concrete aggregate produced from concrete affected by alkali–silica reaction. Cem Concr Res 2010; 40:575–82.

29. Cyr M, Rivard P, Labreque F. Reduction of ASR expansion using powders ground from various sources of reactive aggregates. Cem Concr Compos 2009; 31:438–46.

30. Malvar LJ, Cline GD, Burke D, Rolling R, Sherman TW, Green JL Alkali silicamitigation: state of the art and recommendations. ACI Mater J 2002; 99: 480–9.

31. Duchesne J, Berube MA Long term effectiveness of supplementary cementing materials against ASR. Cem Concr Res 2001; 31:1057–63.

32. Sheata MH, Thomas MDA The effect of fly ash composition on the expansion of concrete due to the alkali–silica reaction. Cem Concr Res 2000; 30:1063–72.

33. http://www.cembureau.be/.

34. Bignozzi MC, Bonduà S. Alternative blended cement with ceramic residues: corrosion resistance. Cem Concr Res 2011; 41:947–54.

35. Giacomini P. World production and consumption of ceramic tiles. Ceram World Rev 2006; 68:58–76.

36. http://www.cerameunie.eu/.

37. Pacheco-Torgal F, Jalali S. Reusing ceramic wastes in concrete. Constr Build Mater 2010; 24:832–8.

38. Cachim P. Mechanical properties of brick aggregate concrete. Constr Build Mater 2009; 23:1292–7.

39. ASTM C1260 Standard test method for potential alkali reactivity of aggregates; 2007.

40. EN 197-1: Cement. Part 1: Composition, specifications and conformity criteria for common cements; 2007.

41. EN 196-1. Methods of testing cement – Part 1: Determination of strength; 2005.

42. EN 12620: Aggregates for concrete; 2006.

43. EN 1015-3. Methods of test for mortar for masonry – Part 3: Determination of consistence of fresh mortar (by flow table); 2007.

44. Park SB, Lee BC, Kim JH. Studies on mechanical properties of concrete containing waste glass aggregate. Cem Concr Res 2004; 34:2181–9.

45. Topcu IB, Canbaz M. Properties of concrete containing waste glass. Cem Concr Res 2004; 34:267–74.

Chapter 7

Carbonation of Ternary Building Cementing Materials

Lucia Fernández-Carrasco[a], D. Torréns-Martín[a], and S. Martínez-Ramírez[b]

[a]Department of Architectural Technology I, Universitat Politècnica de Catalunya, Spain
[b]Instituto de Estructura de la Materia IEM-CSIC, Spain

ABSTRACT

The carbonation processes of ettringite and calcium aluminate hydrates phases developed by hydration of calcium aluminate cement, fly ash and calcium sulphate ternary mixtures have been studied. The hydrated samples were submitted to 4% of CO_2 in a carbonation chamber, and were analysed, previous carbonation and after 14 and 90 days of carbonation time, by infrared spectroscopy and X-ray diffraction; the developed morphology was performed with the 14 days carbonated samples. The results evidenced that ettringite reacts with CO_2 after 14 days of exposition time and evolves totally at 90 days; the developed hydrated phases C_3AH_6 in samples with major CAC content, also reacts with CO_2. Due to carbonation, calcium carbonate – mainly vaterite but also aragonite-, depending on the initial formulation, aluminium hydroxide and gypsum were detected.

INTRODUCTION

The Portland cements-based materials have been widely used in building and civil construction. In the modern architecture these materials often act as the principal constituent in the building, within the structural system or elements of the building envelope – facade, roofing, flooring. Considering the manufacturing process of cement, the average CO_2 emission for the production of each cubic metre of concrete is about 0.2 t [1]; the CO_2 emission from cement manufacture is about 5–8% of the global total attributed to anthropogenic activities [2]. The cement industry and research community are making efforts to improve not only manufacturing processes but also the understanding of cement chemistry to develop sustainable construction materials. At present, energy use, CO_2 emissions and other factors related to sustainability are the main drivers for innovation in cementing materials research. The range of supplementary cementing materials is broadening but also new clinker types are being developed to lower environmental impacts [2]. In this context, Damineli et al. propose indicators for measuring cement use efficiency, present a benchmark based on literature data, and discuss potential gains in cement use efficiency [3].

The supplementary cementing materials, i.e., fly ash, slag, etc., are one of the actual alternatives to reduce consume of large amounts of natural raw materials – clay and calcareous rocks-giving up to a huge spectrum of blended cements that have been noted not only for their properties of reducing energy consumption and CO_2 emission (RECCOE) but also for their well known durability properties [4]. Together with mentioned use of wastes, a recent strategy is the reduction of the clinker factor [5] and then allowing the use millions of tonnes of by-products by reducing the consumption of Portland cement per unit volume of concrete.

Currently, the blend of ordinary Portland cement (OPC), calcium aluminate cement (CAC) and a type of calcium sulphate is of much research interest [6], [7], [8] and [9]. These products based on ettringite have a broad range of uses: formulations with water contents near the minimum requirement to ensure plasticity are widely used in proprietary floor screeds and repair materials. The strength properties depend on formulation but it is comparable with those achieved by sulphate-free CAC cements. In the context of using wastes, new attempts to obtain ettringite-rich materials by the incorporation of fly ash (FA) as an alternative high sustainable ternary systems was investigated obtaining clearly good products [10] and [11].

The first approach on durability of "traditional" ternary systems preformed by Lamberet [12]; however, there exist very little published data on the behaviour of these materials over time; the most work to date on the carbonation of cementitious materials has concentrated on the study of OPC [13]. In OPC, the main developed phases are portlandite and the gel C–S–H. However, it has been reported [14] that under dry conditions, ettringite remained essentially stable, with fine gypsum and vaterite crystals forming rapidly on the surface but an excess of humidity lead to complete dissolution of ettringite to gypsum, vaterite and alumina gel [14] and [15]. But, a crane column exposed to air for 16 years show that ettringite persisted even in very near surface layers; it was suggested that it was a partially carbonate-substituted ettringite and $CaCO_3$ is only formed towards the final stages of carbonation [15].

The goal of this work is to study the evolution of developed AFt phases due to the action of CO_2 developed in the CAC/FA/C$\bar{S}$Hx systems. Six selected formulations of the system were submitted to an accelerated carbonation treatment. The CO_2 incorporation to the ettringite or aluminates structures or decomposition are investigated by infrared spectroscopy, X-ray diffraction and scanning electron microscopy.

MATERIALS, METHODS AND INSTRUMENTAL SET-UP

An Electroland CAC, a fly ash (Type F according to the ASTM classification), and calcium sulphate 2-hydrate (Panreac PRS-CODEX) were used in this research as raw materials. A PHILIPS PW 2400 X-ray fluorescence spectrometer with a PW 2540 VTC sample changer was used to determine the chemical composition of the materials (see Table 1).

Table 1: Chemical composition of raw materials

	LOIa	Al2O3	CaO	SiO2	Fe2O3	MgO	K2O		Na2O
CAC	0.36	40.52	34.89	2.94	12.91	0.5	0.06		0.14
Fly ash	1.27	25.84	5.93	41.49	20.76	1.22	1.37		–

aLOI: loss on ignition.

For the study, the formulations used are given in Table 2 and were prepared in a mixer by blending the raw materials during 1 h at a speed of 9 rpm. The Fig. 1 shows the studied compositions in the ternary system $CAC/FA/CSH_x$ and as the total $CaO/SiO_2/Al_2O_3$ content in the mixtures. The "water/binder" ratio used was 0.6 and the samples were prepared as described in a previous work [11].

Table 2: Portioning of raw materials

System	CAC	FA	C$
S1	20	70	10
S2	30	60	10
S3	40	40	20
S4	40	20	40
S5	80	10	10
S6	90	5	5

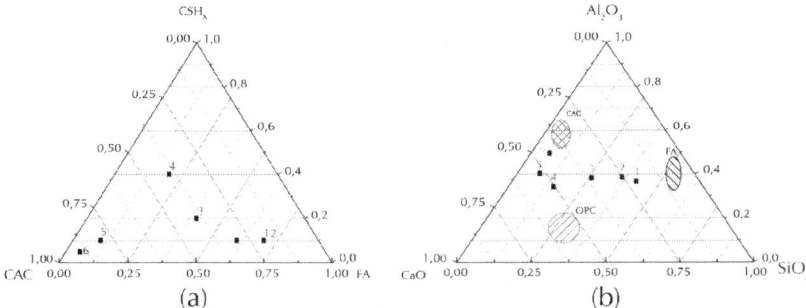

Figure 1: (a) Formulations of the ternary system studied, (b) ternary system of the total $CaO/SiO_2/Al_2O_3$ content.

Carbonation Procedure

After curing, the hydrated pastes were submitted to a carbonation process in a chamber of $50 \times 30 \times 20$ cm³ dimensions. The gas −4% of CO_2 in air – was insufflating during 5 min twice each 24 h; the climatic conditions were maintained at 20 °C of temperature and 75% of relative humidity. The samples were taken after 14 and 90 days of carbonation time and analysed through the different techniques.

The mineralogical composition of the hydrated and carbonated materials was performed by an X-ray diffractometer (XRD) with a Siemens D500 instrument and a Bomem MB-120 Fourier transform infrared spectrophotometer (FTIR) with a frequency range of 4000–450 cm⁻¹ was used. Potassium bromide pellets procedure was used for FTIR analysis. A JEOL JSM-6300 scanning electron microscopy (SEM) and attached a LINK ISIS-200 Energy Dispersive X-ray analysis (EDX) was used to determine the morphology of samples.

RESULTS AND DISCUSSION

Infrared Spectroscopy

Raw materials characterisation has been done in a previous work, for more detailed information see [11]. In the identification of these materials is useful to consider the infrared spectrum of ettringite [16] which presents a very strong anti-symmetrical stretching frequency of the sulphate ion ($_3$ SO_4) centred towards 1120 cm^{-1}; this band is indicative of relative isolation of this ion in the hexagonal prism structure. The water absorption bands appear in the region 1600–1700 cm^{-1} (1640 and 1675 cm^{-1} $_2$ H_2O) and above 3000 cm^{-1} (3420 due to $_1$ H_2O and 3635 cm^{-1} of OH_{free}). The presence of AlO_6 and Al–O–H bending bands are sited near to 550 cm^{-1} and 855 cm^{-1}, respectively.

In the studied systems, the formation of ettringite is explained by the chemical reactions of the sulphates in presence of the aluminates [17] and [18] as described in the following equations:

$$CA + 3C\bar{S}Hx + 2C + (32 - 3x)H \rightarrow C_3A3C\bar{S}H_{32}$$

(1)

$$CA + C\bar{S}Hx + H \rightarrow C_3A3C\bar{S}H_{32} + AH_3$$

(2)

Lamberet [2] in relation with the "traditional" ternary blend CAS-PC-C$\bar{S}$Hx studies of hydration and durability established that the hydration mechanisms depend on PC/CAC and CAC/C$\bar{S}$Hx ratios. The portion of ettringite should be explained in terms of CAC/C$\bar{S}$Hx portions and then it will be directed related to the ratio CAC/C$\bar{S}$Hx added as has been reported [11]. In this research, depending on the formulation the ettringite portion is different; then, also an increase of fly ash on formulations provokes a lower ettringite formation in the hydrated systems. By this reason, not only the presence of ettringite can be explained through the CA/CSH$_x$ ratio.

By the other hand, the CAC rich formulations are on CAC hydrated dominance and calcium aluminates hydrates are formed but also AFm phases due to a deficiency of sulphates. Systems on fly ash formulation dominance would present major ettringite as fly ash content decrease (S4 > S3 > S2 > S1) due mainly to the increase in CA and sulphates available to reacts.

The inspection of the IR spectra of noncarbonated S1–S4 samples show clearly the ettringite presence throughout the bands at 3638, and 3430 cm^{-1} and near to 1665, 1110, 988 and 855 cm^{-1}; other absorption bands near to 600 and 536 cm^{-1} are assigned also to ettringite. The IR spectrum of S1 and S3 samples presented a very slight absorption bands near to 1025 and 990 cm^{-1} indicating AH$_3$ presence. Moreover, the IR spectrum of S4 sample presented absorption bands due to gypsum, at 1690, 1620 and 670 cm^{-1}and the absorption band towards to 600 cm^{-1} is also due to the contribution of gypsum.

With major CAC content, the IR spectra of systems S5 and S6 are almost identical; but a weak band near to 3665 cm^{-1} characteristic of OH$_{free}$ from cubic calcium aluminate hydrates, C$_3$AH$_6$ is observed in the S6 spectrum. Both IR spectrums presented absorption bands towards 3620, 3525 and 3465 cm^{-1} due to aluminium hydroxide as gibbsite. An absorption band towards 3684 cm^{-1} (-OH$_{free}$) could indicate presence of C$_4$AH$_{13}$[19] and also an absorption band near 3640 cm^{-1} would indicate ettringite; and a band around 3673 cm^{-1} could be due to calcium monocarboaluminate. Most important difference between S5 and S6 spectrums are the relative intensity of the absorption band sited near 1110 cm^{-1} – of major relative intensity in S5 sample spectrum; an anti-symmetrical stretching frequency of the sulphate ion ($_3$-SO$_4$) indicates the relative isolation of this ion in the hexagonal prism, although a greater degeneracy of the symmetry would result in more bands in this area of the spectrum of ettringite. The absorption bands near to 1025 and 970 cm^{-1} are assigned to AH$_3$. At lower frequencies, the aluminate bands towards 790 cm^{-1}(Al–O–H bending) and 530 cm^{-1} (-AlO$_6$) are not suitable for identification because C$_3$AH$_6$, monosulfoaluminate, C$_4$AH$_{13}$ and ettringite present very close absorption bands.

Due to carbonation, the analysis of IR spectra of 14 days carbonated samples show a very broad absorption area between 3650 towards 3350 cm^{-1} with maximums near to 3621, 3525 and 3467 cm^{-1} due mainly to aluminium hydroxide as gibbsite (Fig. 2). The spectra show a broad strong $_3$ carbonate bands between 1600 and 1300 cm^{-1} – centred near 1480 cm^{-1} in S1, S2, S5 and S6 spectra but towards to 1495 cm^{-1} in S3 and S4 spectra, respectively. The $_1$ –CO$_3$$^{2-}$ bands appear near 876 cm^{-1} in the spectra of S1, S2, S5 and S6 samples but close to 855 cm^{-1} in the S3 and S4 spectra. The developed calcium carbonate polymorph was vaterite and aragonite, respectively. In the spectra was also observed the $_4$carbonate bands at 713 and 700 cm^{-1}, clearly in the S3, S4, S5 and S6 spectra and as a shoulder for S1 and S2 spectra. These results indicate that the main calcium carbonate polymorph formed is vaterite in samples S1, S2; aragonite is developed through carbonation of S3 and S4 samples and both polymorphs in S5 and S6.

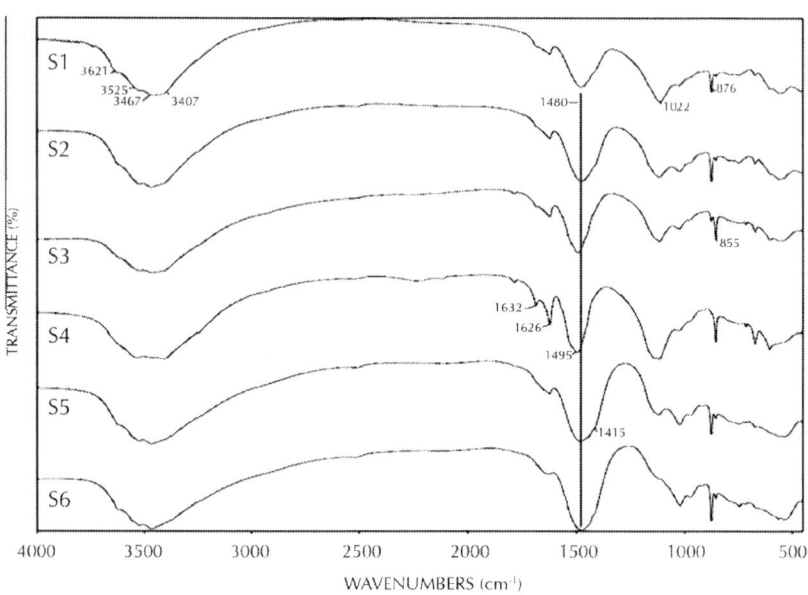

Figure 2: IR spectras of samples at 14 carbonation days.

At 90 days of carbonation, S1–S4 spectra also show a very broad absorption area between 3650 towards 3350 cm^{-1} but some modifications can be observed in the absorption band position of maximums with respect to the spectra at 14 carbonation days: 3540, 3470, 3410 and 3245 cm^{-1} (Fig. 3). However, the absorption bands due to AH_3 increase in spectra of S5 and S6 samples, with maximums at 3624, 3525 and 3468 cm^{-1}. The absorption bands due to carbonate compounds do not experiment an important modification – S1, S2, S5 and S6 present absorption bands at 1480 and 875 cm^{-1} while samples S3 and S4 at 1500 and 855 cm^{-1}, indicating the presence of same carbonate compounds –vaterite and aragonite, respectively; but they are higher in intensity at 90 days carbonation time showing a major portion of them (Fig. 3).

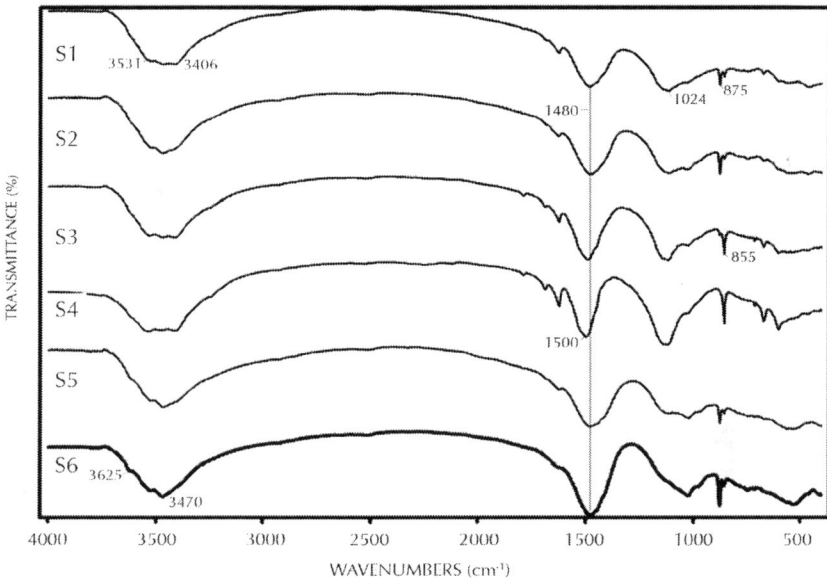

Figure 3: IR spectras of samples at 90 carbonation days.

The position of the absorption bands sited near to 3540, 3470, 3410, 3245, 1685, 1623, 1143, 1115, 670 and 600 cm^{-1} indicate the presence of a sulphate compound similar to gypsum. The intensity of the mentioned bands indicate major portion in S4 sample and then S3, minor in S1, S2 and S5, respectively (Fig. 3). Moreover, the presence of 1025 cm^{-1} absorption band, indicating the aluminium hydroxide compounds presence, is specially noted in the spectrums of S6 and then S5 samples; but they are present in all spectra.

It is interesting to remark that there were not observed intense bands of ettringite both in 14 but also in the 90 days carbonated spectrum, this will be discussed after the X-ray diffraction analysis of these samples.

X-Ray Diffraction

The main crystalline hydration product of the reactions in the S1–S4 studied systems was ettringite. Apart from ettringite, the sample S4 presented a high portion of nonreacted gypsum. The samples S5 and S6 – with major portion of CAC- presented minor ettringite content but in addition also presented C_3AH_6; and the broad diffraction lines between $2 = 6-16°$ indicated calcium monosulfoaluminate and hemicarboaluminate. The diffraction lines for aluminium hydroxide polymorphs are observed only in the diffraction patterns of samples with high CAC content (S5 and S6). Moreover, the samples S5 and S6 presented a very low portion of the compound C_2ASH_8.

Due to the chemical reactions of them with CO_2, an evolution of hydrated phases towards the formation of calcium carbonate, gypsum and aluminium hydroxide is observed. The Table 3 presents a semiquantitative interpretation made of the analysis of the major developed hydrated and carbonated compounds detected in the diffraction patterns. The Fig. 4, Fig. 5 and Fig. 6 presented diffraction patterns of hydrated and carbonated samples – 14 and 90 days, respectively.

Table 3: Major identified compounds by XRD

Sample	Curing time	Ettringite	Gypsum	Calcite	Aragonite	Vaterite	AH3
S1	Hydrated	1.3	0	0	0	0	0
	14 days	1	0.7	0	0	1.3	0.5
	90 days	0	10	0	12	13	0
S2	Hydrated	1	0	0	0	0	0
	14 days	0.5	0.7	0	0	1.6	0.5
	90 days	0	9	0	12	15	0
S3	Hydrated	3	0	0	0	0	0
	14 days	2	2	0	2	0	1
	90 days	0	16	0	12	0	0
S4	Hydrated	4	2	0	0	0	0
	14 days	1	6	0	1.4	0	0.5
	90 days	0	50	0	11	0	0
S5	Hydrated	0.5	0	0	0	0	0.4
	14 days	0.2	0.3	1	0.8	2	1
	90 days	0	7	14	0	20	9
S6	Hydrated	0.3	4	0	0	0	0.4
	14 days	0	0	0	0	1.7	1
	90 days	0	0	0	0	12	10

The number relates to number of measured counts (1–100 counts).

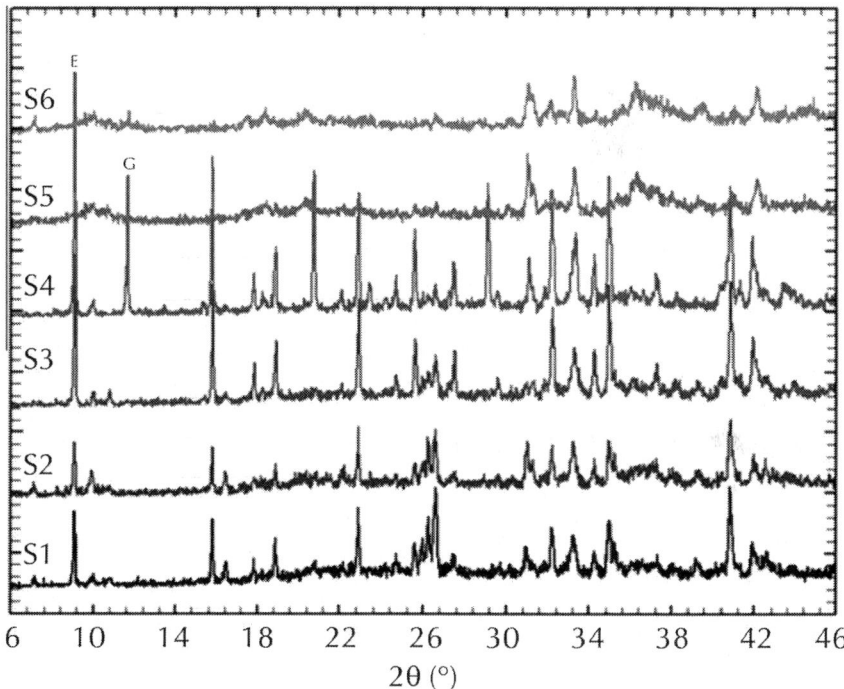

Figure 4: X-ray diffraction patterns of hydrated samples.

At 14 carbonation days, the relative intensity of diffraction lines of ettringite decrease and no signals of it were shown at 90 carbonation days. Other minor hydration products, i.e., C_3AH_6, calcium monosulfoaluminate, hemicarboaluminate and C_2ASH_8, also reacted totally with CO_2.

The crystalline detected products were calcium carbonate, aluminium hydroxide and gypsum. At 14 carbonation days, the main detected calcium carbonate polymorph was vaterite; but the samples S3 and S4 presented aragonite in major portion than vaterite. In the pattern of sample S5 were observed less intense diffraction lines due to calcite and aragonite (Fig. 5). At 90 days, an important increase in the presence of vaterite was observed for samples S1, S2, S5 and S6. Samples S1 and S2 also increased in aragonite. By the other hand, the main calcium carbonate polymorph detected at 90 days carbonation time for samples S3 and S4 was aragonite (Fig. 6).

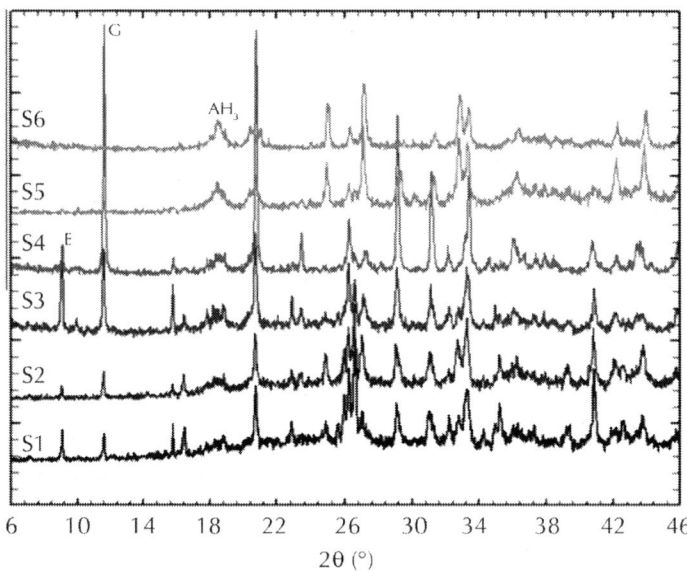

Figure 5: X-ray diffraction patterns of samples at 14 days of carbonation.

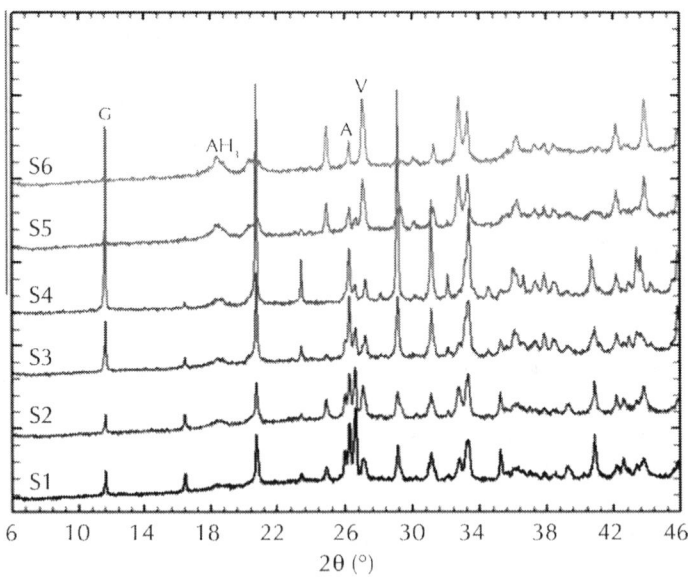

Figure 6: X-ray diffraction patterns of samples at 90 days of carbonation.

Zhou and Glasser [15] studied ettringite stability under controlled relative humidity at 68 and 88% in different solid forms, as powder and in pellets. Ettringite completely transformed to gypsum and vaterite, when it was in powder form, while only the exposed surface presented complete dissolution in pellets. In the exposed pellets, the layers beneath revealed partial dissolution and transformation to monophase, while in the deepest layer, ettringite remained intact. In the present study – carbonation of solids as powder and at 75% of RH – a different developed calcium carbonate was analysed depending on the formulation: samples S1, S2, S5 and S6 presented mainly vaterite; but samples S3 and S4 presented important amounts of aragonite. The carbonation of samples also produced the formation of aluminium hydroxide and gypsum – the formation of that sulphate type compound, probably gypsum was also detected by XRD in carbonated samples S1–S5 but mainly in S3 and S4; however more research needs to be done on this aspect in order to determine possible modifications on its crystal structure. In such context, it has been reported about the existence of a carbonate analogue of SO_4-AFt as CO_3-AFt, but the synthesis of this compound was presented with some difficulties. Poellman et al. [20] and Barnett et al. [21] indicated the existence of possible solid solutions formed between SO_4-AFt and CO_3-AFt. Moreover, Peng et al. [22]explained that the ettringite itself undergoes union exchanges in which part of its sulphate is replaced by carbonate; they suggest that ettringite undergoes significant ion exchange, of CO_3 for SO_4, prior to its decomposition and thus also buffers the paste against carbonate penetration. In the studied samples and with the analytical used techniques is not possible to confirm these ion exchanges but only a suspicious can be mentioned due to the modification of the position of certain absorption bands in the infrared or in the diffraction lines in the XRD patterns.

Respecting the aluminium hydroxide, at 14 carbonation days, its diffraction lines were observed in the all pattern of samples –S1 to S6. However, at 90 carbonation days, aluminium hydroxide was only detected as a crystalline phase in pattern of S5 and S6 samples. The formation of a sulphate type compound, probably gypsum was also detected by XRD at 14 carbonation days in samples S1–S5; however more research need to be done on this aspect. Major portion of this sulphate was detected at 90 days, mainly in samples S4 and S3 (Fig. 5 and Fig. 6). Another aspect to be in present is that during the hydration process of cements if waste ions are mixed with, several types of waste ion interactions may occur in the cement microstructure: chemisorb, precipitate, form a surface compound to any of several cement phases surfaces, form inclusions or be chemically incorporated into the cement structures, or have simultaneous occurrence of several of these situations. Then, ettringite structure can withstand modest deviations in composition without a change in structure. This compositional change occurs on the crystal chemical level in the form of ionic substitution; the ions available for substitution in the ettringite structure are Ca^{+2}, Al^{3+}, SO_4^{2-} and OH^-. The examination of ettringite group structures provides information on some natural and, therefore, stable ionic substitutions possible in the structure [20]. The aluminium site accommodates a variety of trivalent and tetravalent cations and this can be a place of instability respecting the CO_2.

Scanning Electron Microscopy

Most important results from the morphological analysis developed of 14 days carbonated samples were the observed differentiate morphology of calcium carbonates. The calcium carbonate morphology of samples S1, S2, S5 and S6 was very poor and aggregate, with a very low particle size (Fig. 7). On the other hand, the microstructure of calcium carbonate developed in samples S3 and S4 was observed as needles of around 2 μm size (Fig. 8). No ettringite crystals were identified at this age but spheres of remained fly ash can also be observed intermixed and normally surrounded by films or crystals products growing from their surface.

Figure 7: Dense morphology of carbonates.

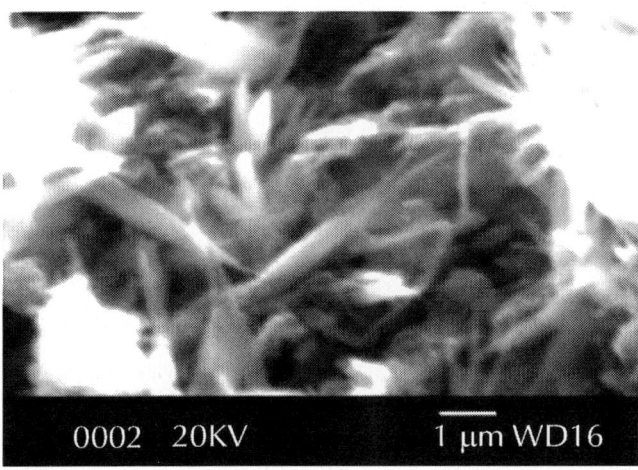

Figure 8: Needle-like morphology of carbonates.

The revised results indicated that samples S1, S2, S5 and S6 presented, at 14 days, only calcium carbonate as vaterite while samples S3 and S4 the calcium carbonate at the same age was developed as aragonite, then it can be justified the different morphologies of carbonates – the vaterite is developed as aggregates with very a poor morphology while aragonite is develop as needles in this systems.

Many researchers have expressed concern about the long term stability and indicated a rapid rate of carbonation of ettringite-based matrices [15] and [23]. Carbon dioxide would dissolve ettringite not only due to the inferior pH value due to absence of portlandite in these systems, but also through direct carbonation reactions. The morphology of pastes –and specially that of ettringite would have a notable importance on durability. The special properties of these systems differentiated kinetics and giving up to a rapid set and hard times. Obviously, the kinetics will affect the morphology of hydrated developed phases and a conexion could be found with the developed carbonated phases but more research need to be done in this aspect. In a recent work [24] formation of metastable phases of calcium carbonate has been correlated with inorganic substrates, i.e., aluminium oxide promote vaterite formation. Additionally particles size can also have effect in metastable calcium carbonate formation.

CONCLUSIONS

The following conclusions can be drawn from the present work:

Sample exposure to CO_2 affects the evolution of the hydrated phases, due to the chemical reactions of them with CO_2, towards the formation of calcium carbonate, gypsum and aluminium hydroxide. It has been reported that the carbonation of the "traditional" ettringite (i.e., without ionic substitutions in the structure due to wastes) evolves towards calcium carbonate and gypsum [14].

A different developed calcium carbonate was analysed depending on the initial CAC/fly/ash/C$\overline{S}$Hx formulation: samples S1, S2, S5 and S6 presented mainly vaterite; but samples S3 and S4 presented important amounts of aragonite. As has been reported in the bibliography the carbonation of ettringite has been associated to vaterite.

At early age carbonation, aluminium hydroxide was detected in the diffraction patterns of all the samples. However, at 90 carbonation days, aluminium hydroxide was only detected in samples S5 and S6. Then, secondary chemical reactions due to AH_3 presence take place: reaction due to carbonation develops new phases.

The morphology of calcium carbonate of samples S3 and S4 appeared as needle-like and is attributed to aragonite, while calcium carbonate from samples of S1, S2, S5 and S6 appeared as an aggregate dense morphology and attributed to vaterite.

ACKNOWLEDGEMENTS

The authors thank The Ministerio de Economía y Competitividad (Mineco) for the BIA00767-2008 project.

REFERENCES

1.   Price L, Worrell E, Phylipsen D. Energy use and carbon dioxide emissions in energy-intensive industries in key developing countries. In: Proceeding of the 1999 earth technologies forum. Washington DC; September 1999. p. 27–29.

2.   Scrivener Karen L, James Kirkpatrick R. Innovation in use and research on cementitious materials. Cem Concr Res 2008;38:128–36.

3.   Damineli BL, Kemeid FM, Aguiar PS, Vanderley MJ. Measuring the ecoefficiency of cement use. Cem Concr Compos 2010;32(8):555–62.

4.   Macphee DE. Sustainable cementitious binders: new chemistries, new performance. Advanced in cement and concrete X. Sustainability. ECI Engineering conferences international. Davos (Switzerland); 2006.

5.   Bensted J, Barnes P. Structure and performance of cements. 2nd ed. New York: Spon Press; 2002.

6.   Chatterji S, Jeffery JW. Studies of early stages of paste hydration of cement compounds: II. J Am Ceram Soc 1963;46 (4):187–91.

7.   Brooks SA, Sharp JH. In: Manghabhai RJ, editor. Calcium aluminate cements ettringite-based cements. London: E. & F. Spon; 1990. p. 335–49.

8.   Lamberet S. Durability of ternary binders based on Portland cement, calcium aluminate cement and calcium sulphate. These no. 3151. Switzerland: Ècole Polytechnique Fédérale de Lasanne (EPFL); 2005.

9.   Fernandez-Carrasco L. Formación de etringita en mezclas ternarias. X Cong Nacion Mater Santander 2008;II:979–82.

10.   Fernandez-Carrasco L, Martínez-Ramírez S. Infrared and Raman spectroscopy analysis of products from CAC–fly ash–CSHx hydrated mixtures. In: The Fred Glasser Cement Science Symposium. University of Aberdeen; 2009. p. 75–6.

11.   Fernández-Carrasco L. Reactions of fly ash with calcium aluminate cement and calcium sulphate. Fuel 2009;88:1533–8.

12.   Lamberet S. 1st calcium aluminate cement congress. Avinyon: The Centenary Congress; 2008.

13.   Rostami V, Shao Y, Boyd AJ, He Z. Microstructure of cement paste subject to early carbonation curing. Cem Concr Res 2012;42:186–93.

14.   Glasser FP. The stability of ettringite. In: Proceedings of rilem meeting on delayed ettringite formation. Villars-sur-Ollon, Switzerland; September 4–6, 2002.

15.   Zhou Q, Glasser FP. Thermal stability and decomposition mechanisms of ettringite at <120 C. Cem Concr Res 2001;31:1333–9.

16.   Bensted J. An infrared spectral examination of calcium aluminate hydrates and calcium aluminate sulphate hydrates encountered in Portland cement hydration. Italy: Alluminati di Calcio, Seminario Internazionale Torino; 1982.

17.   Taylor HFW. Cement Chemistry. Thomas Telford; 1997.

18.   Chatterji S, Jeffery JW. Studies of early stages of paste hydration of cement compounds: II. J Am Ceram Soc 1963;4 6(4):187–91.

19.   Volant J. Etude par spectrométrie infrarouge d´ aluminates de calcium hydratés. Thèse de doctorat, Publication du centre d´ études et de recherches de l´industrie des Liants Hydrauliques; 1966.

20.   Poellman H, Kuzel JH, Wenda R. Solid solution of ettringites part I: incorporation of OH and CO_2 3 in $3CaOAl_2O_332H_2O$. Cem Concr Res 1990;20:941–7.

21.   Barnett SJ, Adam CD, Jackson ARW. An XRPD profile fitting investigation of the solid solution between ettringite, $Ca6Al2(SO_4)3(OH)1226H_2O$ and carbonate ettringite $Ca_6Al_2(CO_3)3(OH)1226H_2O$. Cem Concr Res 2001;31:13–7.

22.   Peng JH, Zhang. The mechanism of the formation and transformation of ettringite. J Wuhan Univ Technol-Mater Sci Edit 2006;21(3):158–61.

23.   Matschei T, Lothenbach B, Glasser FP. Thermodynamic properties of Portland cement hydrates in the system $CaO–Al_2O_3–SiO_2–CaCO_3–H_2O$. Cem Concr Res 2007;37:1379–410.

24.   Lin Y, Hu Q, Chen J, Ji J, Teng HH. Formation of metastable $CaCO_3$ polymorphs in the presence of oxides and silicates. Crystal Growth Des 2009;9(11):4634–41.

Experiments and Predictions of Physical Properties of Sand Cemented by Enzymatically-induced Carbonate Precipitation

Hideaki Yasuhara[a], Debendra Neupane[a], Kazuyuki Hayashi[b], and Mitsu Okamura[a]

[a]Department of Civil and Environmental Engineering, Ehime University, Matsuyama 790-8577, Japan

[b]Department of Civil Engineering, Wakayama National College of Technology, 77, Nadacho Noshima, Gobo 644-0023, Japan

ABSTRACT

A grouting technique that utilizes precipitated calcium carbonate as a cementing material is presented. The enzyme urease is used to enhance the rate and the magnitude of the calcium carbonate precipitation.

Evolutions in the mechanical and the hydraulic properties of treated sand samples are examined through unconfined compression and permeability tests, respectively. The grout is mainly composed of urease, which bio-catalyzes the hydrolysis of urea into carbon dioxide and ammonia, urea, and calcium chloride solutions. This method employs chemical reactions catalyzed by the enzyme, and ultimately acquires precipitated calcium carbonate within soils. The mechanical test results show that even a small percentage of calcium carbonate, precipitated within soils of interest, brings about a drastic improvement in the strength of the soils compared to that of untreated soils—the unconfined compressive strength of the samples treated with <10 vol% calcium carbonate precipitation against the initial pore volume ranges from ~400 kPa to 1.6 MPa. Likewise, the hydraulic test results indicate the significant impervious effects of the grouting technique—the permeability of the improved samples shows more than one order of magnitude smaller than that of the untreated soils. Evolutions in the measured hydraulic conductivity and porosity are followed by a flow simulator that accounts for the solute transport process of the injected solutions and the chemical reaction of the calcite precipitation. Predictions of the changes in permeability with time overestimate the test measurements, but those of the changes in porosity show a good agreement with the actual measurements, indicating that such simulations should become a significant supplementary tool when considering real site applications.

INTRODUCTION

Chemical grouting has been used occasionally as a countermeasure against the liquefaction of the ground beneath existing structures. Recently, a novel grouting method that utilizes precipitated calcium carbonate as a cementing material has been examined. The precipitation of calcium carbonate is induced by the microbial metabolism (e.g.,Stocks-Fischer et al., 1999, Nemati et al., 2005 and DeJong et al., 2006). This technology, using the microbial metabolism, may be effective for stiffening soils of interest and for reducing the permeability of the soils. However, the evolutions in the mechanical and the hydraulic properties induced by the microbial metabolism may not be straightforward enough to be controlled, because it may be impossible to constrain the extinction and/or the generation of living bacteria in natural environments.

The research on calcium carbonate precipitation by bacteria has been mainly conducted using ureolytic bacteria. These bacteria indirectly produce precipitated calcium carbonate by a urease enzyme. The bacterium selected for research on calcium carbonate precipitation, containing the urease enzyme, is typically *Sporosarcina pasteurii*. The microbial-induced carbonate precipitation (MICP) has been evaluated as a soil-strengthening process, concluding that MICP has the potential to improve the mechanical properties of porous materials on a typical sample scale (Le Metayer-Levrel et al., 1999, Nemati and Voordouw, 2003, DeJong et al., 2006, Whiffin et al., 2007 and Sugimoto and Kuwano, 2009), and on a larger container scale (Van der Ruyt and van der Zon, 2009 and van Paassen et al., 2010). In this MICP technique, the transport and the fixation of the bacteria of interest are significant issues for achieving a suitable level of improvement of the saturated porous media, and thus, have been studied to this end (Murphy and Ginn, 2000, Foppen and Schijven, 2006, Whiffin et al., 2007 and Harkes et al., 2010). In contradiction to the rigorous experimental works on this topic, the theoretical and/or numerical works are sparse; the rate of the microbially-induced urea hydrolysis has been evaluated (Fujita et al., 2008), but research on the prediction of evolutions in the mechanical and/or the hydraulic properties of the improved materials induced by the MICP is not apparent.

In this work, the urease enzyme is adopted instead of using bacteria such as *Sporosarcina pasteurii*, often used as a promoter for the hydrolysis of urea, which, as described above, causes Ca^{2+} and CO_3^{2-} to precipitate as $CaCO_3$ and form into the void spaces and/or the surfaces of grains. Utilizing the enzyme itself is more straightforward than using bacteria, because the cultivation and fixation of bacteria (i.e., biological treatment) do not need to be considered in this work. A grouting technique, in which chemically-precipitated calcite is adopted as the cementing material, may be recognized as the calcite in situ precipitation system (CIPS) (Kucharski et al., 1996, Ismail et al., 2002a and Ismail et al., 2002b). The CIPS may be similar to the method that is being presented in this work. However, the CIPS is a commercial product and the composition of its chemical solution is not apparent. In this work, the compositions of the adopted solutions are clearly addressed. After introducing the grouting reagents into the soil samples, the evolutions in the mechanical and the hydraulic properties are examined through unconfined compression tests and permeability tests, respectively. Moreover, changes in the hydraulic conductivity are predicted by an advection–diffusion simulation by considering the calcium carbonate precipitation, and the predictions are compared with the actual measurements.

EXPERIMENTS

In order to examine the workability of the grout materials utilized in this work, test-tube experiments are conducted. Then, a suite of unconfined compression and permeability experiments is conducted for the calcium carbonate-precipitated sand under various initial and boundary conditions by changing the amount of reactants. The reactants used and the experiments conducted here are explained in detail.

Materials

Urease (Kishida Chemical Co., Ltd.,: 020-83242) is found in bacteria and in several plants, such as sword beans, and is used throughout the current work as an enzyme to hydrolyze urea. The resulting carbonate ions are applied to produce the calcium carbonate being precipitated. The companion to calcium carbonate (i.e., calcium ions) is supplied from the calcium chloride solution, which is freely soluble in water (i.e., the solubility is 82.8 g/100 mL of water at 20 °C). The expected reactions to obtain the calcium carbonate precipitation, enhanced by the effect of urease, are expressed as follows:

$$CO(NH_2)_2 + 2H_2O \rightarrow 2NH_4^+ + CO_3^{2-}$$

(1)

$$CaCl_2 \rightarrow Ca^{2+} + 2Cl^-$$

(2)

$$Ca^{2+} + CO_3^{2-} \rightarrow CaCO_3 \downarrow$$

(3)

where $CO(NH_2)_2$ represents urea. A schematic of the whole process listed above and the grouting mechanism expected are illustrated in Fig. 1.

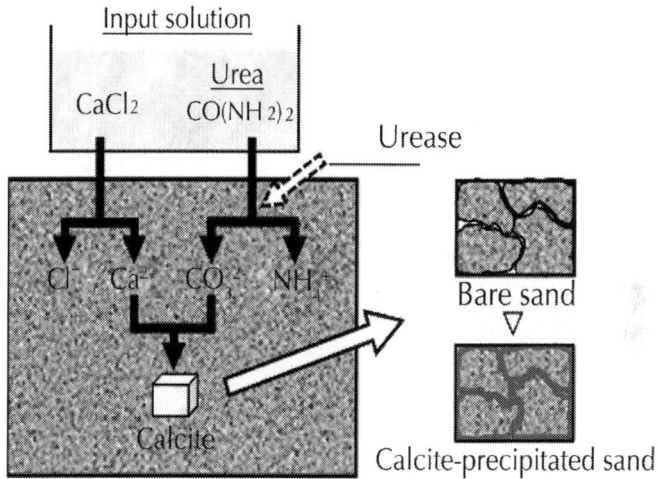

Figure 1: Schematic of calcium carbonate-precipitation process and grouting mechanism.

Test-tube Experiments

In this work, urease, urea, and calcium chloride are used as reagents contained in the grout. The rate and the magnitude of calcium carbonate precipitation should be controlled by the amounts of those materials. Thus, the effects of the grouting materials exerted on the calcium carbonate precipitation are examined by conducting two different sorts of test-tube experiments. A schematic of the test-tube experiments is shown in Fig. 2. During the experiments, the solutions in the test tubes are stirred regularly by a rotating table to ensure complete mixing.

Test-tubes

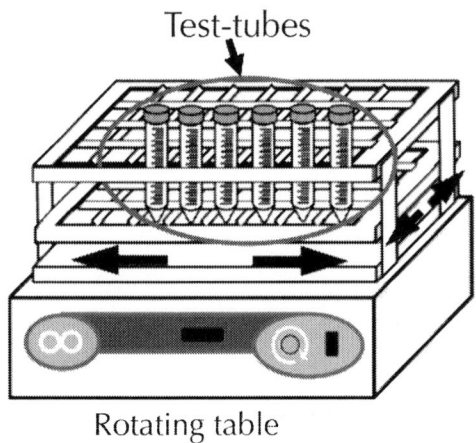

Rotating table

Figure 2: Schematic of test-tube experiments.

The aim of one set of test-tube experiments is to examine the rate of urea hydrolysis that is bio-catalyzed by the urease. When urea is dissociated into ammonium and carbonate ions (Eq. (1)), the pH of the solutions should increase correspondingly to the production of ammonium ions. Therefore, the measurements of the evolution in pH with time may indirectly define the rates and the magnitude of the urea dissociation accelerated by the urease. Note that the rate of the changes in pH resulting from the production of ammonium ions is not equivalent to that of the calcium carbonate precipitation, and that the precipitation should take more time than the dissociation of urea. The experimental conditions for this purpose are listed in Table 1. As shown in the table, the concentration of urea solutions is fixed at 0.5 mol/L, while the amounts of urease are varied. The evolving pH is measured by a pH meter (KRK: KP-5F) at 0, 1, 2, 3, 6, 12, and 24 h after mixing.

Table 1: Experimental conditions of test-tube experiments

Sample	Amount of urease [g/100 mL of solution]	Urea concentration [mol/L]	CaCl$_2$ concentration [mol/L]
TpH-1	0.5	0.5	0
TpH-2	1.0		
TpH-3	1.5		

TCa-1	1.0	0.5	0.5
TCa-2		1.0	1.0
TCa-3		1.5	1.5

The aim of the other set of test-tube experiments is to examine the calcium carbonate-precipitation characteristics depending on the concentrations of grout materials. The experimental conditions for this purpose are also listed in Table 1. As shown in the table, the amount of urease is fixed at 1 g/100 mL of water, while the concentrations of $CaCl_2$–urea solutions are varied. The solutions sampled 24 h after mixing were assayed by inductively-coupled plasma atomic emission spectrometry (ICP-AES) to quantify the Ca concentrations.

The results of the two different test-tube experiments are shown in Fig. 3. As is shown inFig. 3(a), all measured pH levels increased rapidly at 1 h, and then approached a steady state after 6 h, although the case of TpH-1 showed a slight increase at 12 h. The peak pH for cases TpH-1, 2, and 3 ranged from 9.53 to 9.62, which shows no prominent difference among the experimental conditions adopted here. Fig. 3(b) represents the relation between the initial prescribed Ca concentrations and the consumed Ca concentrations that are equivalent to the initial values minus the measurements by ICP-AES. The average Ca consumption ratios to the initial values of TCa-1, 2, and 3 are 96.4, 98.7, and 79.1%, respectively. These results indicate that the high $CaCl_2$–urea concentrations relative to the amount of urease may restrain the activity of urease, which may in turn result in a reduction in the calcium carbonate precipitation.

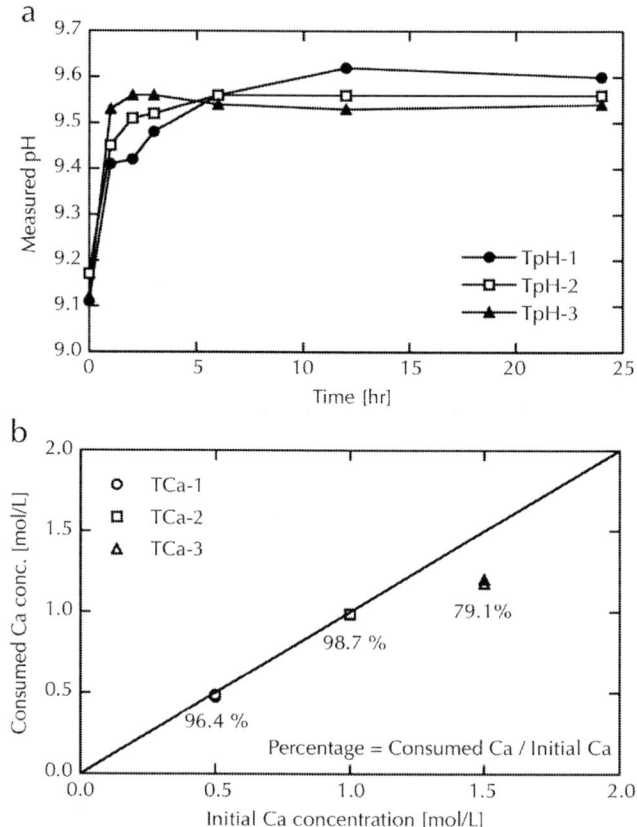

Figure 3: Results of test-tube experiments ((a) changes in pH with time and (b) relation between initial Ca and consumed Ca concentrations).

The test-tube experiments have revealed both the rapid dissociation rates of urea, accelerated by the urease, and the importance of the relative concentrations between the urease and the $CaCl_2$–urea solutions.

Unconfined Compression Tests

In this section, unconfined compression tests are conducted to examine the effects of the improvement exerted on the stiffness and the strength of treated sand samples. The procedure to prepare the samples for the tests is as follows. The test apparatus is shown in Fig. 4. Firstly, 300 g of dry Toyoura sand, well-mixed with a certain amount of urease powder, is carefully pluviated in air to acquire a relative density of 50% (i.e., an initial porosity of 0.44) with sample dimensions of 50 mm in diameter and 100 mm in height (roughly 290 g of mixed sand is consumed in the sample preparation). The urease powder is pre-mixed with the sand in order to achieve homogeneous samples. Secondly, after the dry sample is evacuated to facilitate saturation, as the $CaCl_2$–urea solution is to be injected, a confining pressure of 50 kPa is applied. Thirdly, a concentration-fixed calcium chloride solution, blended with the same molar urea, is injected into the dry sand samples at the prescribed times, and then the samples are cured for 24 h within a pressure cell. Fourthly, 150 mL of distilled water is injected to flush out the byproducts of chloride and ammonium ions. Finally, the cured sand samples are taken out of the cell and are dried completely). The dried samples are used for the unconfined compression tests.

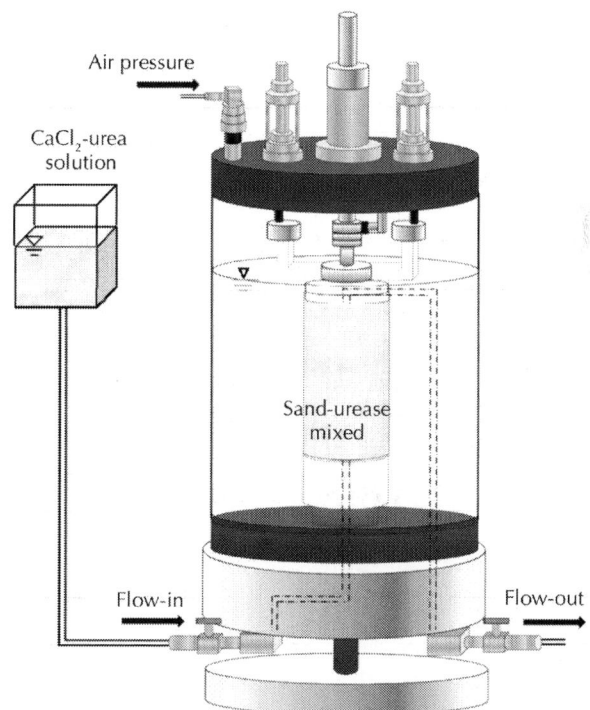

Figure 4: Pressure cell for sample preparation.

The test conditions for the unconfined compression tests are listed in Table 2. The concentrations of the $CaCl_2$–urea solution are 0.5 and 1.0 mol/L—the same molar concentrations for $CaCl_2$ and urea are blended, in advance. The amounts of urease, mixed well with 300 g of Toyoura sand, are 0.5 and 1.0 g. One hundred milliliters of the $CaCl_2$–urea solutions are injected into the samples (i.e., sand+urease) over approximately 0.5 h, and the same amount of solutions is injected 4 or 8 times at 2-h intervals. Here, a maximum amount of calcium carbonate (i.e., calcite) precipitation is estimated to be 40 g (i.e., 0.4 mol) for every test where the reactions in Eqs. (1),(2) and (3) fully proceed. To check the reproducibility, two samples are made for each test condition.

Table 2: Experimental conditions for unconfined compression experiments

Sample case	Sample name	Concentration [mol/L][a]	Injection number	Urease [g][b]
C1	bio-1, 4, 6	0.5	8	1.0
C2	bio-2, 5	1.0	4	1.0
C3	bio-3, 7	0.5	8	0.5

[a]Concentrations of $CaCl_2$–urea solutions.

[b]Urease is mixed with 300 g of the Toyoura sand.

Prior to the unconfined compression tests, the improved sand sample of bio-1 (Fig. 5), that is not used for the compression tests, was examined by an X-ray diffractometry (XRD: Rigaku RINT 2200) and a scanning electron microscopy (SEM: Hitachi S-2700 SEM). The sand sample was saw-cut and several specimens were prepared for both analyses. Fig. 6 shows representative XRD results for the bare sand and the improved sand samples. As is most clearly shown in Fig. 6(b), a distinct peak of calcite is observed in the improved sand, guaranteeing that calcite is the precipitated material within the pore spaces that have been induced by the solution injections. Fig. 7 shows the SEM results for the pre- and the post-improved samples. The precipitated materials are the size of a few tens of microns, and are situated on the free-surface and at the boundaries of the grains. These materials are most likely to be calcite, as is clear from the XRD results. The precipitated calcium carbonate, complicatedly stuck around the grains, should result in the manifestation and the augmentation of stiffness and strength.

Figure 5: Improved sand sample of bio-1.

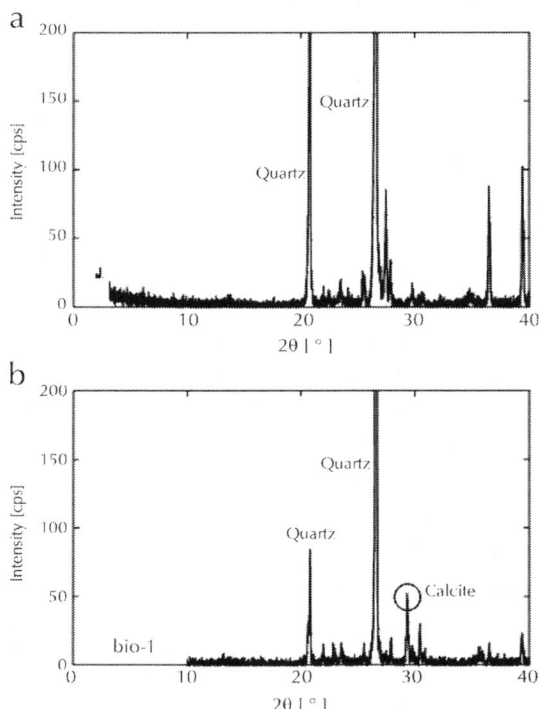

Figure 6: XRD results of (a) bare sand and (b) improved sand (bio-1).

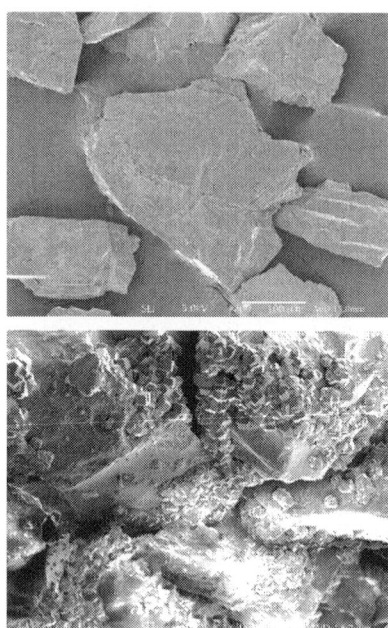

Figure 7: SEM results of bare sand and improved sand (bio-1).

Using the other improved sand samples of bio-2 to bio-7, unconfined compression tests have been conducted. Fig. 8 shows the observed relations between the normal strain and the normal stress, and the secant elastic modulus at 50% of the peak strength (E_{50}) and the unconfined compressive strength (UCS) are evaluated from the observation (Table 3). In Fig. 8, the early behavior does not show a nearly linear trend, but a positive curvature. This is likely to be attributed to a slight sliding along the grain boundaries and the compression of micro-pores, and should be typical behavior for rock materials (e.g.,Jaeger et al., 2007). As is apparent, the observed ranges for E_{50} and UCS of ~50 MPa to ~160 MPa and ~400 kPa to ~1.6 MPa, respectively, are, roughly speaking, comparable to those of weak rocks. The results have revealed that this calcium carbonate precipitation method adequately solidifies loose sand. The bio-5 (or C2) results show the strongest values among all the conditions, while the bio-3 (or C3) results are the weakest. It is understood that the concentrations of the solutions injected and the amounts of urease enhancing the reaction are key parameters to controlling the stiffness and the strength of the improved samples.

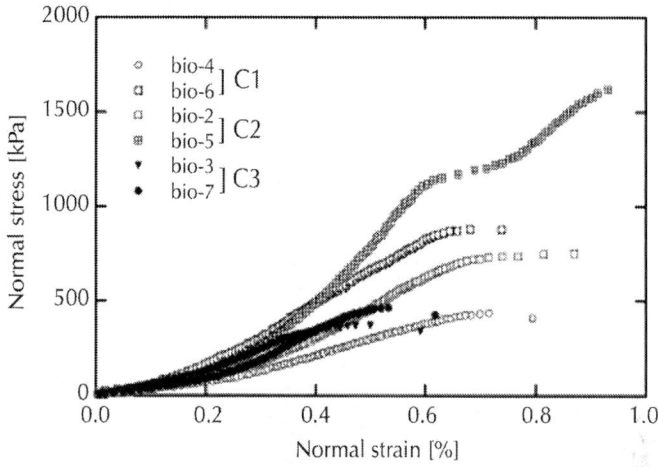

Figure 8: Obtained results of relation between normal strain and normal stress.

Table 3: Evaluated E_{50} and *UCS* from unconfined compression experiments

Sample case	Sample	εf [%] [a]	E_{50} [MPa]	*UCS* [kPa]
C1	bio-4	0.706	53.5	435
	bio-6	0.676	120	890
C2	bio-2	0.870	84.1	754
	bio-5	0.918	160	1620
C3	bio-3	0.481	73.5	373
	bio-7	0.514	71.5	466

[a]Compressive strain at peak stress.

Although a maximum precipitation of calcium carbonate is estimated at 40 g for all cases, the actual amounts of precipitated calcium carbonate need to be examined. For this purpose, after completing the compression tests, the samples were rinsed using a 0.1 M HCl solution three times to dissolve the precipitated calcium carbonate. Then, the disaggregated samples were dried again, and the amounts of calcium carbonate were evaluated by comparing the weights of the pre- and the post-rinsed samples. A relation between the amount of calcium carbonate and *UCS* is depicted in Fig. 9. It is clear from the figure that the actually-precipitated amounts range from 30 to 60 wt% against the maximum. Although some exceptions are observed, there is a tendency for more strength to be observed with a larger amount of calcium carbonate, which is reasonable. The broken lines in the figure represent 5 and 10% of the initial pore volume. Thus, the *UCS* of the improved sand may lie between 400 and 1600 kPa where 5–10% of the pore volume is occupied by precipitated calcium carbonate. In order to compare the results in this study with data from literature, a relation between the amount of calcium carbonate and *UCS*, which is evaluated using the results shown in van Paassen et al. (2010), is depicted in Fig. 10. Since the ratio of calcium carbonate to the weight of dry sand is shown for reference, the amounts are estimated by assuming that the weight of the sand sample is 300 g, which is equivalent to that used in this study. As shown in the figure, the results in this study may follow a regression curve evaluated from those by van Paassen et al. (2010), and may be compatible with those existing data. Taken together, the results of the unconfined compression tests indicate that the *UCS* of the sand improved by this injection method may be controlled when the amount of actual precipitation has been well predicted.

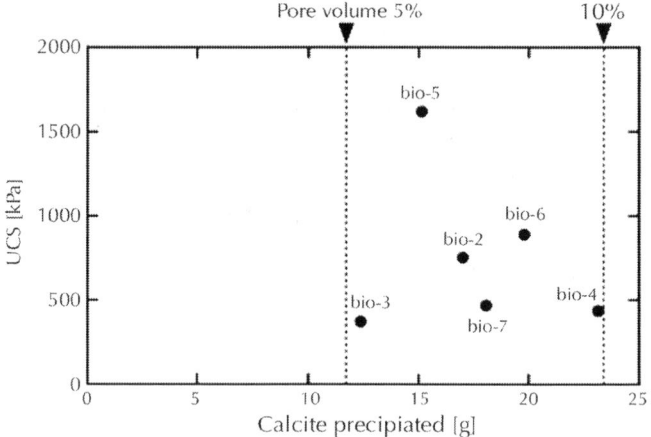

Figure 9: Relation between precipitated calcium carbonate and *UCS*. Broken lines represent 5 and 10% of initial pore volume.

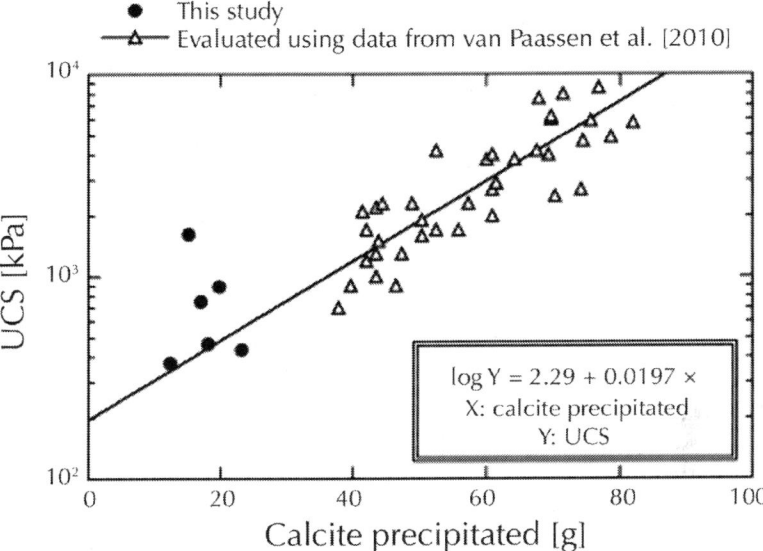

Figure 10: Relation between precipitated calcium carbonate and *UCS*. Open triangles are evaluated using data fromvan Paassen et al. (2010) (S5).

Permeability Tests

In this section, permeability tests are conducted to examine the effects of improvement on the permeability of the treated sand samples. The procedure for the sample preparation for the permeability tests (Fig. 11) is similar in part to that for the unconfined compression tests explained above, but also different in part. It is the same in terms of the amount of Toyoura sand, urease, urea, calcium chloride, and the dimension of the samples. It is different in that the mixed sand is pluviated into an acrylic cylinder to acquire a relative density of 50%, and that the permeability tests are conducted at prescribed intervals after injecting the solutions. Note that no confining pressures are applied throughout the tests.

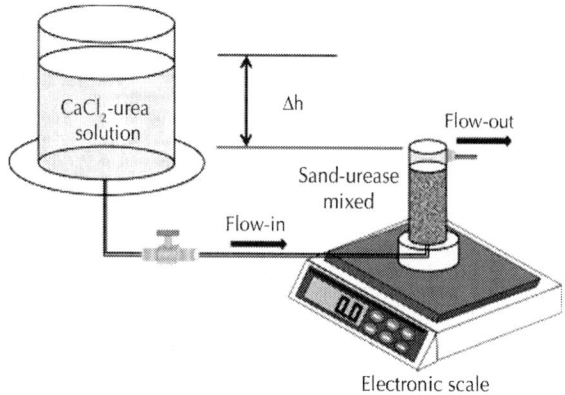

Figure 11: Schematic of permeability tests.

The test conditions for the permeability tests are listed in Table 4. The concentrations of the $CaCl_2$–urea solution are 0.5 and 1.0 mol/L, which are equivalent to those of the unconfined compression tests. The amounts of urease are 1.0 and 2.0 g against 300 g of sand. One hundred and fifty milliliters of $CaCl_2$–urea solutions are injected into the samples over approximately 0.5 h, and the same amount of solutions is injected 1 or 4 times at 2-h intervals. After the completion of the final injection, the hydraulic conductivity is measured under a constant-head condition at 0, 1, 2, 3, 6, 12, and 24 h. In a series of tests, the six different conditions seen in Table 4 are adopted. The number of solution injections is just one for P1–P4, and four for P5 and P6.

Table 4: Experimental conditions for permeability experiments

Sample case	Concentration [mol/L][a]	Injection number	Urease [g][b]	Maximum precipitation [g]
P1	0.5	1	1.0	7.5
P2	1.0	1	1.0	15.0
P3	0.5	1	2.0	7.5
P4	1.0	1	2.0	15.0
P5	0.5	4	1.0	30.0
P6	1.0	4	2.0	60.0

[a]Concentrations of $CaCl_2$–urea solutions.

[b]Urease is mixed with 300 g of the Toyoura sand.

The temporal changes in hydraulic conductivity, observed for P1–P4, are shown in Fig. 12(a). The initial hydraulic conductivity of ~0.04 cm/s decreases monotonically by 60–70% for all cases; it reaches a steady state within roughly 5 h. The monotonic decrease in hydraulic conductivity should be attributed to the calcium carbonate precipitation that occurs and gradually increases after the solution injection. The ultimate values for P1 to P4 are 0.019, 0.012, 0.011, and 0.010 cm/s, respectively. The results show that the molar concentrations of the injected solution (i.e., 0.5 or 1.0 mol/L) and the amount of urease (i.e., 1.0 or 2.0 g/300 g of sand) do not significantly influence the changes in permeability when the number of solution injections is just one.

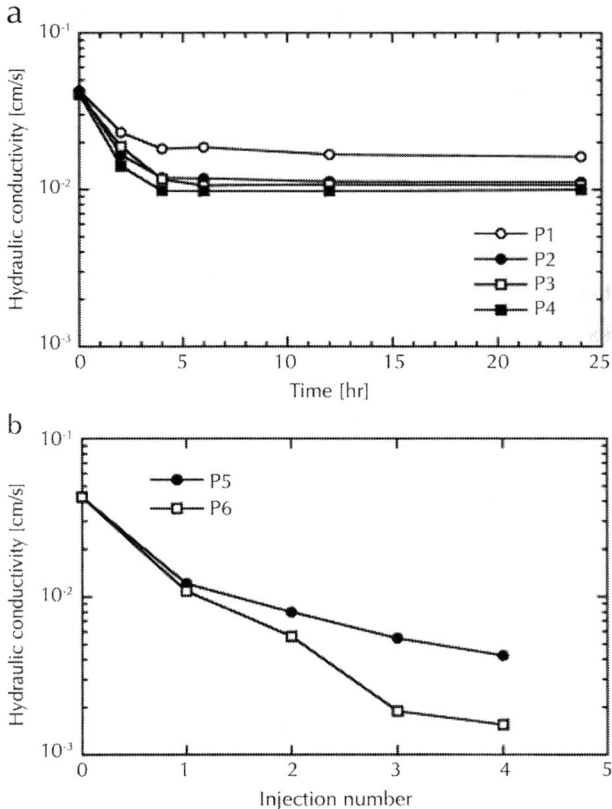

Figure 12: Evolution of hydraulic conductivity ((a) temporal changes for P1 to P4 and (b) different injection numbers for P5 and P6).

In contrast, when the injection number is increased, the decrease in hydraulic conductivity 24 h after the injections is apparent (Fig. 12(b))—the more injection numbers, the more decreases are observed. The ultimate hydraulic conductivity (i.e., 4.2×10^{-3} and 1.5×10^{-3} cm/s for P5 and P6, respectively) decreases by roughly one order of magnitude after the fourth injection, as compared to the initial values. Moreover, the decrease under the higher concentration condition (P6) is greater than that under the lower condition (P5), which is reasonable because more precipitated calcium carbonate may clog more pore spaces within the sample. The decrease measured in the permeability tests is relatively significant (ca. one order of magnitude reduction), but is compatible to the decreases measured in the permeability tests conducted using the samples improved by the similar bio-grouting technique (Nemati and Voordouw, 2003 and Kawasaki et al., 2006).

As examined in the previous section, the amount of actually precipitated calcium carbonate is also evaluated through the above-mentioned acid leaching. The relation between the precipitated calcium carbonate and the observed hydraulic conductivity is depicted in Fig. 13; it shows an obvious log-linear tendency between them. This indicates that the permeability of the sand improved by this injection method may be controlled when the amount of actual precipitation has been well predicted.

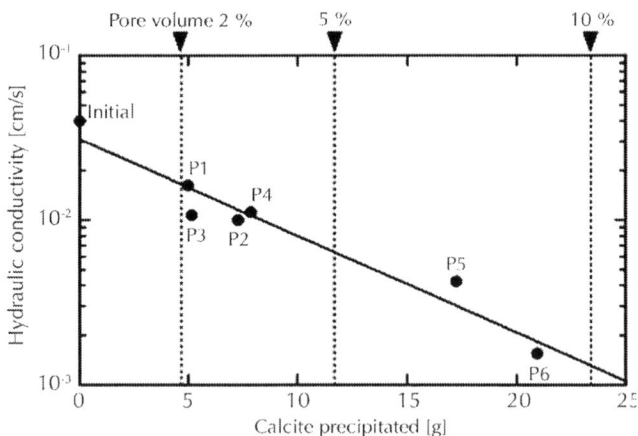

Figure 13: Relation between precipitated calcium carbonate and hydraulic conductivity. Broken lines represent 2, 5 and 10% of initial pore volume.

MODEL PREDICTIONS

During injections of the grouting materials proposed in this work, an advective and dispersive transport of the solutions, together with chemical reactions of the calcite precipitation, occur within the pore spaces of the targeted soils. Thus, the process of the chemical reactions should be solved in a model that takes into account the advection–dispersion process. In this work, a suite of mathematical equations, used for the transport simulations, is presented. Then, a comparison is made of the results between the measurements of the permeability tests and the predictions.

Mathematical Formation for Solution Transport with Chemical Reactions

In this work, a non-isothermal reactive geochemical transport code, TOUGHREACT (Xu et al., 2004), is utilized to follow the evolution in permeability resulting from calcium carbonate precipitation mediated by enzyme-driven mineralization. Therefore, a whole calculation procedure is explained in detail by Xu et al. (2004). In this section, a summary of the calculation adopted in this work is explained.

Calcite precipitation is only considered for chemical reactions in this work. The governing equation is the advection–diffusion equation with chemical reactions, namely,

$$\frac{\partial(\phi C_j)}{\partial t} = -\nabla(u C_j) + \tau \phi D \nabla^2 C_j + R_n,$$

(4)

where ϕ [–] is the porosity and C_j [mol/m^3] is the concentration of aqueous chemical component j. u [m/s] is the Darcy velocity, τ [–] is the medium tortuosity, and D [m^2/s] is the diffusion coefficient. R_n [mol/m^3/s] is the reaction term (i.e., calcium carbonate precipitation), which is expressed as

$$R_n = -k_n A_n \rho_w \left| 1 - \Omega_n^{\theta} \right|^{\eta},$$

(5)

where k_n [mol/m^2/s] is the precipitation rate constant, A_n [m^2/kg] is the specific reactive surface area per kg H_2O, ρ_w [kg/m^3] is the water density, and Ω_n [–] is the kinetic mineral saturation ratio. Parameters θ and η are the constants that are constrained from the dissolution experiments. In this work, the measured BET specific surface area is adopted as An. For calcium carbonate precipitation, these parameters are considered unity. The rate constant is typically defined by

$$k_n = k_{25}\exp\left[\frac{-E_a}{R}\left(\frac{1}{T} - \frac{1}{298.15}\right)\right],$$

(6)

where k_{25} represents the rate constant at 25 °C, E_a [J/mol^1] is the activation energy, R[J/K/mol] is the gas constant, and T [K] is the absolute temperature.

The evolution of the porosity induced by the calcite precipitation is followed with time simply by evaluating the volume precipitated as

$$\phi(t+\Delta t) = (1 + R_n \times V_m \times \Delta t)\phi(t),$$

(7)

where t [s] is the time step in the calculations and Vm [m^3/mol] is the molar volume (i.e., 3.69×10^{-5} m^3/mol for calcite). The changes in permeability are calculated from the changes in porosity using the Carman–Kozeny relation (Bear, 1972), which ignores the changes in grain size, tortuosity, and the specific surface area, given by

$$K(t) = K_0\frac{(1-\phi_0)^2}{(1-\phi(t))^2}\left(\frac{\phi(t)}{\phi_0}\right)^3,$$

(8)

where K [m^2] is the permeability and subscript "0" represents the initial condition. Permeability may be converted to hydraulic conductivity, which is familiar to civil engineers and given by

$$k(t) = K(t)\frac{\rho_w g}{\mu},$$

(9)

where k [m/s] is the hydraulic conductivity, g [m/s^2] is the gravity, and μ [Pa s] is the dynamic viscosity of the water. Specifically, an effective hydraulic conductivity in the vertical direction is evaluated to be compared with those obtained from the constant-head permeability tests, clearly defined as

$$\overline{k}_v(t) = \frac{h}{\sum(h_i/k_i(t))},$$

(10)

where $\overline{k}_v(t)$ is the effective hydraulic conductivity in the vertical direction, h [m] is the sample height (i.e., 0.1 m), h_i [m] is the height of element i, and k_i [m/s] is the hydraulic conductivity of element i.

Comparison between Measurements and Predictions

To simulate the circumstances occurring in the permeability tests, such as the hydrolysis of urea by urease, a certain amount of carbonate and ammonium ions are made to exist in the calculation domain as initial conditions in the analysis. Then, 150 mL solutions of calcium and chloride ions, whose concentrations are equivalent to those used in the permeability tests, are injected into the domain, whose dimensions are equivalent to those of the test samples (i.e., 50 mm in diameter and 100 mm in height). Subsequently, a curing environment is simulated, after the solution injection, by flowing de-mineralized water at an extremely slow flow rate (i.e., $\sim 10^{-30}$ kg/s), whose manipulation is adopted because calculations of the chemical reactions are not executed under no-flow conditions. The parameters (Xu et al., 2004 and Barkouki et al., 2011), and the initial and boundary conditions used for the analysis, are listed in Table 5 and Table 6, respectively.

Table 5: Parameters used for numerical analysis

Parameter	Value
Grain diameter, d	210 µm
Particle density, ps	2.64 g/cm^3
Temperature, T	20 °C
Initial porosity, φ	0.44
Specific surface area, An	0.98 m^2/kg[a]

Initial hydraulic conductivity, K_0	4.4×10^{-2} cm/s
Precipitation rate constant, k_{25}	1.0×10^{-8} mol/m²/s[b]
Activation energy, Ea	62.8 kJ/mol[a]

[a]An and Ea are obtained from Xu et al. (2004).

[b]k_{25} is obtained from Barkouki et al. (2011).

Table 6: Initial and boundary conditions for numerical analysis

	Injection period (0–0.5 h)		Curing period (0.5–24 h)	
	Initial conditions	Boundary conditions	Initial conditions	Boundary conditions
Concentrations of Ca^{2+}, Cl^-, HCO_3^-, NH_4^+ [mol/L]	1.0×10^{-9}	0.50	<0.50[a]	1.0×10^{-9}
Flow rate [kg/s]	–	8.33×10^{-5}	–	8.33×10^{-30}

[a]The remaining concentrations after consumed during the injection period are prescribed as the initial conditions.

A comparison of the results between the predictions and the test measurements for P1 and P5 is shown in Fig. 14. As seen in the figure, the predictions significantly overestimate the actual values. This may be attributed to the conversion equation from porosity to permeability (i.e., Eq. (8)). Permeability in porous media is strongly dependent upon the grain size (or the related pore size and its distribution), but it is not considered in this equation. Moreover, the precipitation occurring within the test samples may not be fully homogeneous. If a relatively large amount is precipitated locally, this may impede water flow, resulting in a significantly greater reduction in permeability than that predicted. Therefore, this analysis is incapable of following the changes in grain or pore size and its distribution, which should occur in the process of calcite precipitation.

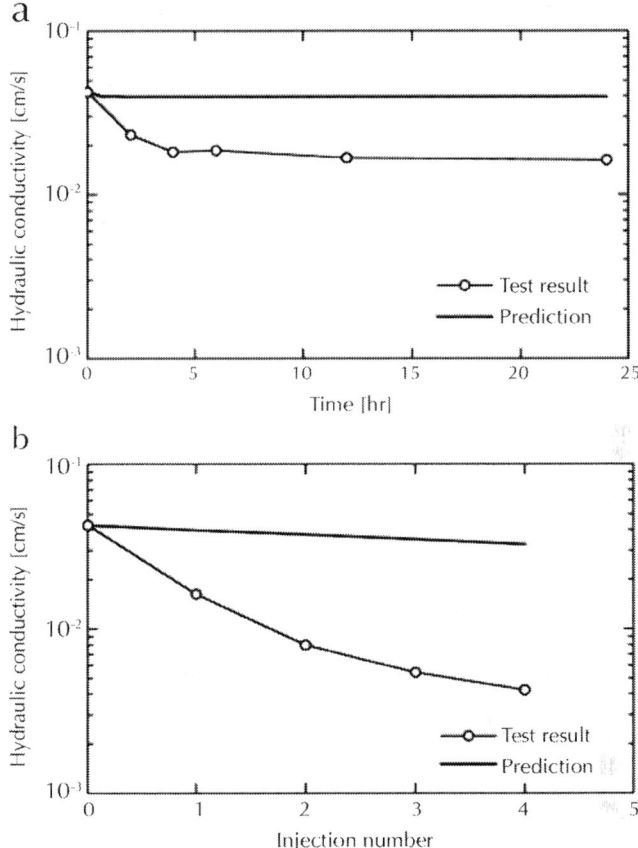

Figure 14: Comparison of changes in hydraulic conductivity between test results and predictions for P1 and P5.

Alternatively, the changes in porosity are compared between the predictions and the actual measurements. The porosity of the post-test samples is evaluated by adopting the method explained in Section 2.4 (i.e., weighing the amount of precipitated calcium carbonate by means of an acid leaching). The results are compared in Fig. 15. As shown in the figure, the predictions match the measurements for P1 and P5 quantitatively well. Moreover, by observing Fig. 14(a) and Fig. 15(a), it should be noted that the predictions follow the experimental behavior quantitatively well—the predicted porosity monotonically decreases and reaches a steady state around 5 h, which is congruent with the evolution of the measured hydraulic conductivity. As one may imagine, the predicted reduction in porosity should be controlled by an adopted calcite precipitation rate constant. Two different values of k_{25} for calcite are obtained from the literature—6.46×10^{-7} mol/m²/s from Xu et al. (2004) and 3.81×10^{-7} mol/m²/s from Nilsson and Sternbeck (1999). These values are slightly greater than that shown in Table 5. The predictions using k_{25} of 3.81×10^{-7} mol/m²/s are also shown in Fig. 15. As expected, the rate in the early period and the reduction in the magnitude of porosity are slightly greater than those obtained using the original value shown in Table 5, but are still compatible with the test results.

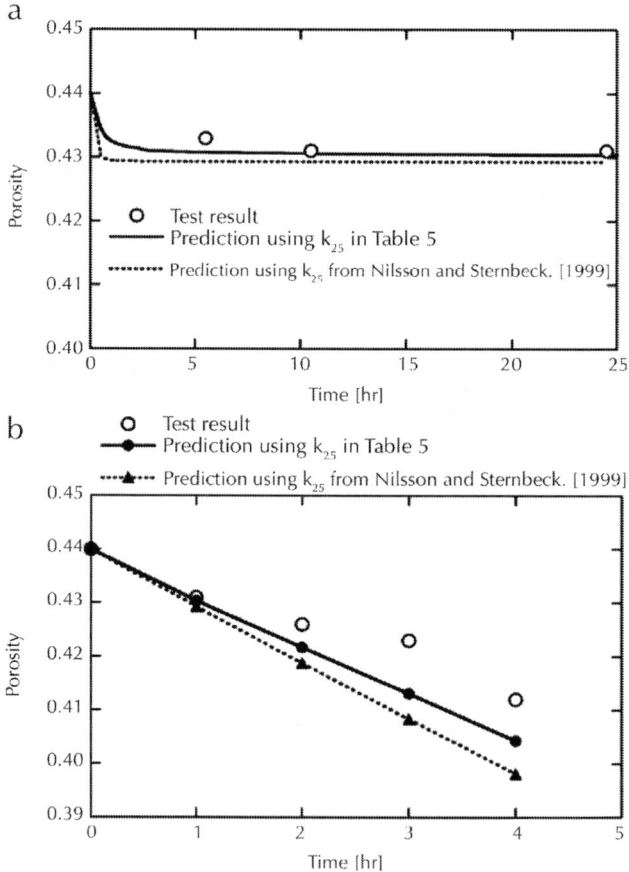

Figure 15: Comparison of changes in porosity between test results and predictions for P1 and P5.

Fig. 16 represents the changes in the porosity distribution with time under the P1 conditions. Before the solution injection, the initial porosity of 0.44 is prescribed in the whole domain. Then, the porosity decreases gradually with time from the bottom, as the curing period proceeds. After 6 h of curing, no conspicuous difference (i.e., a further decrease in porosity) is observed. Ultimately, a slight discrepancy in the porosity is apparent between the top and the bottom elements— the porosities predicted at the top and the bottom are 0.438 and 0.425, respectively. After the permeability tests under the P1 conditions, the porosity was actually evaluated at the top, the middle, and the bottom of the sample. The obtained results at the top, the middle, and the bottom are 0.430, 0.432, and 0.432, respectively, which are qualitatively congruent with the predictions.

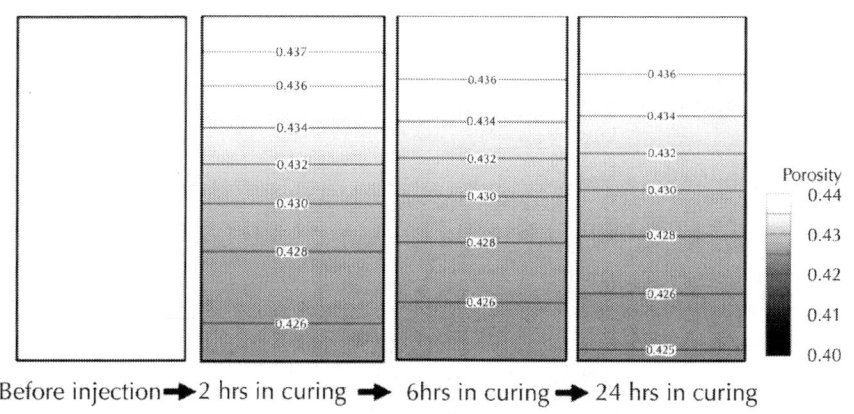

Before injection➡2 hrs in curing ➡ 6hrs in curing ➡ 24 hrs in curing

Figure 16: Evolution of porosity distribution with time under P1 conditions.

The porosity discrepancy between the top and the bottom is attributed to the slow flow rate prescribed in this work, which is 150 mL/30 min, while an initial pore volume of the soil sample is roughly 85 cm^3. Since the advection of the targeted ions of Ca^{2+} and CO$_3^{2-}$ is slow, relative to the chemical reaction of the calcite precipitation, the targeted ions are consumed readily at the upper stream; and consequently, a discrepancy in porosity within the whole domain occurs. This result implies that a relative relation between advection (or related flow rates) and chemical reactions of the calcite precipitation should be fully examined and identified in advance whenever this grouting technique is to be applied at real sites.

The above outcomes lead to the conclusion that the numerical model should be effective for simulating the rate and the magnitude of the evolutions in porosity induced by calcite precipitation, and should be an important tool that supplements the grouting technique presented in this work. In order to achieve better predictions, proper relations between porosity and permeability, which take into account changes in grain size, tortuosity, and specific surface area, should be obtained.

CONCLUSIONS

This work has experimentally and numerically examined a grouting technique that utilizes calcium carbonate precipitation mediated by enzyme-driven mineralization. Unconfined compression and permeability tests, conducted for the improved samples, have shown the efficacy of the technique. Specifically, the stiffened samples are produced – unconfined compression strength ranges from 400 kPa to 1.6 MPa and the impervious properties are achieved – the permeability of the improved samples is reduced by more than one order of magnitude. Both the compression and the permeability test measurements implicate that the strength and the permeability of the improved body can be constrained when the amount of calcium carbonate being precipitated by this technique are well controlled, although further tests are needed.

Numerical analyses that simulate the advection–diffusion process, complemented by consideration of the chemical reactions of the calcite-precipitation kinetics, are conducted to replicate the changes in permeability measured in the permeability tests. The predictions significantly overestimate the actual measurements of permeability, but show a good agreement with the changes in porosity measured. This means that this model may be applicable to the prediction of the rates and the magnitudes of soil improvement resulting from the current grouting technique.

The grouting technique presented in this work, using the enzyme itself, may be more straightforward than that using bacteria which generates the enzyme of interest, because one can skip the process of cultivating the bacteria. The current grouting technique, in which the enzyme is mixed well within the soil samples prior to injecting the solutions, should be unsatisfactory. A solution containing the enzyme should be injected into the samples from the outside. This methodology will be examined and reported in the near future.

ACKNOWLEDGMENTS

This work has been partly supported by research grants from the Institute for Fermentation, Osaka, Japan and the Shikoku Kensetsu Kousaikai. Their support is gratefully acknowledged. The authors also thank Mr. Tomohiro Sugimoto and Mr. Koichi Kado for their help with the experimental portion of this study.

REFERENCES

1.   Barkouki, T.H., Martinez, B.C., Mortensen, B.M., Weathers, T.S., De Jong, J.D., Ginn, T.R., Spycher, N.F., Smith, R.W., Fujita, Y., 2011. Forward and inverse bio-geochemical modeling of microbially induced calcite precipitation in half-meter column experiments. Transport in Porous Media.

2.   Bear, J., 1972. Dynamics of Fluids in Porous Media. Dover Publications, Inc. (pp. 764).

3.   DeJong, J.T., Fritzges, M.B., Nu¨sslein, K., 2006. Microbially induced cementation to control sand response to undrained shear. Journal of Geotechnical and Geoenvironmental Engineering, ASCE 132 (11), 1381–1392.

4.   Foppen, J.W.A., Schijven, J.F., 2006. Evaluation of data from the literature on the transport and survival of Escherichia coli and thermotolerant coliforms in aquifers under saturated conditions. Water Research 40, 401–426.

5.   Fujita, Y., Taylor, J.L., Gresham, T.L.T., Delwiche, M.E., Colwell, F.S., Mcling, T.L., Petzke, L.M., Smith, R.W., 2008. Stimulation of microbial urea hydrolysis in groundwater to enhance calcite precipitation. Environmental Science & Technology 42, 3025–3032.

6.   Harkes, M.P., van Paassen, L.E., Booster, J.L., Whiffin, V.S., van Loosdrecht, M.C.M., 2010. Fixation and distribution of bacterial

activity in sand to induce carbonate precipitation for ground reinforcement. Ecological Engineering 36, 112–117.

7. Ismail, M.A., Joer, H.A., Randolph, M.F., Meritt, A., 2002a. Cementation of porous materials using calcite. Geotechnique 52, 313–324.

8. Ismail, M.A., Joer, H.A., Sim, W.H., Randolph, M.F., 2002b. Effect of cement type on shear behavior of cemented calcareous soil. Journal of Geotechnical and Geoenvironmental Engineering 128, 520–529.

9. Jaeger, J.C., Cook, N.G.W., Zimmerman, R.W., 2007. Fundamentals of Rock Mechanics, 4th ed. Blackwell Publishing (pp. 475).

10. Kawasaki, S., Murao, A., Hiroyoshi, N., Tsunekawa, M., Kaneko, K., 2006. Fundamental study on novel gout cementing due to microbial metabolism. Journal of the Japan Society of Engineering Geology 47, 2–12.

11. Kucharski, E., Price, G., Li, H., Joer, H.A., 1996. Engineering properties of CIPS-cemented calcareous sand. In: Proceedings of the 30th International Geological Congress. Beijing, Brill Academic. pp. 92–97.

12. Le Metayer-Levrel, G., Castanier, S., Orial, G., Loubiere, J.F., Perthuisot, J.P., 1999. Applications of bacterial carbonatogenesis to the protection and regeneration of limestones in buildings and historic patrimony. Sedimentary Geology 126, 25–34.

13. Murphy, E.M., Ginn, T.R., 2000. Modeling microbial processes in porous media. Hydrogeology Journal 8, 142–158.

14. Nemati, M., Voordouw, G., 2003. Modification of porous media permeability, using calcium carbonate produced enzymatically in situ. Enzyme and Microbial Technology 33, 635–642.

15. Nemati, M., Greene, E.A., Voordouw, G., 2005. Permeability profile modification using bacterially formed calcium carbonate: comparison with enzymic option. Process Biochemistry 33, 925–933.

16. Nilsson, O., Sternbeck, J., 1999. A mechanistic model for calcite crystal growth using surface specification. Geochimica et Cosmochimica Acta 63, 217–225.

17. Stocks-Fischer, S., Galinat, J.K., Bang, S.S., 1999. Microbiological precipitation of CaCO3. Soil Biology and Biochemistry 31, 1563–1571.

18. Sugimoto, D., Kuwano, R., 2008. Trial test on the evaluation of soil cementation generated by the function of microorganism. In: Proceedings of the 7th International Symposium on New Technologies for Urban Safety of Mega Cities in Asia. October 2008, USMCA, Beijing.

19. pp. 669–674.

20. Van der Ruyt, M., van der Zon, W., 2009. Biological in situ reinforcement of sand in near-shore areas. Geotechnical Engineering 162, 81–83.

21. Van Paassen, L.A., Ghose, R., van der Linden, T.J.M., van der Star, W.R.L., van Loosdrecht, M.C.M., 2010. Quantifying biomediated ground improvement by ureolysis: large-scale biogrout experiment. Journal of Geotechnical and Geoenvironmental Engineering 136, 1721–1728.

22. Whiffin, V.S., Van Paassen, L.A., Harkes, M.P., 2007. Microbial carbonate precipitation as a soil improvement technique. Geomicrobiology Journal 24 (5), 417–423.

23. Xu, T., Sonnenthal, E., Spycher, N., Pruess, K., 2004. TOUGHREACT User's Guide: A Simulation Program for Non-Isothermal Multiphase Reactive Geochemical Transport in Variably Saturated Geologic Media. Lawrence Berkeley Lab. Report LBNL-55460. Berkeley, California. pp. 192.

Effect of Tricalcium Aluminate on the Physicochemical Properties, Bioactivity, and Biocompatibility of Partially Stabilized Cements

Kai-Chun Chang[1], Chia-Chieh Chang[1], Ying-Chieh Huang[1], Min-Hua Chen[2], Feng-Huei Lin[2], and Chun-Pin Lin[1]

[1]Graduate Institute of Clinical Dentistry, School of Dentistry and National Taiwan University Hospital, National Taiwan University, Taipei, Taiwan,

[2]Institute of Biomedical Engineering, National Taiwan University, Taipei, Taiwan

ABSTRACT

Background/Purpose

Mineral Trioxide Aggregate (MTA) was widely used as a root-end filling material and for vital pulp therapy. A significant disadvantage to MTA is the prolonged setting time has limited the application in endodontic treatments. This study examined the physicochemical properties and biological performance of novel partially stabilized cements (PSCs) prepared to address some of the drawbacks of MTA, without causing any change in biological properties. PSC has a great potential as the vital pulp therapy material in dentistry.

Methods

This study examined three experimental groups consisting of samples that were fabricated using sol-gel processes in C3S/C3A molar ratios of 9/1, 7/3, and 5/5 (denoted as PSC-91, PSC-73, and PSC-55, respectively). The comparison group consisted of MTA samples. The setting times, pH variation, compressive strength, morphology, and phase composition of hydration products and ex vivo bioactivity were evaluated. Moreover, biocompatibility was assessed by using lactate dehydrogenase to determine the cytotoxicity and a cell proliferation (WST-1) assay kit to determine cell viability. Mineralization was evaluated using Alizarin Red S staining.

Results

Crystalline phases, which were determined using X-ray diffraction analysis, confirmed that the C3A contents of the material powder differed. The initial setting times of PSC-73 and PSC-55 ranged between 15 and 25 min; these values are significantly ($p<0.05$, ANOVA and post-hoc test) lower than those obtained for MTA (165 min) and PSC-91 (80.5 min). All of the PSCs exhibited ex vivo bioactivity when immersed in simulated body fluid. The biocompatibility results for all of the tested cements were as favorable as those of the negative control, except for PSC-55, which exhibited mild cytotoxicity.

Conclusion

PSC-91 is a favorable material for vital pulp therapy because it exhibits optimal compressive strength, a short setting time, and high biocompatibility and bioactivity.

INTRODUCTION

The pulpal tissue in teeth has various functions, including (1) physiological deposition of dentin by odontoblasts; (2) nutritional supply through microcirculation; (3) protection and sensation of nerve endings in dentin, and (4) repair under stimulation by forming tertiary dentin. [1] When pulpal tissue is subjected to trauma, caries, or iatrogenesis, the tooth can be treated with root canal treatment or pulp capping to prevent further pulpal inflammation or infection. Although the success rate of root canal treatment is high, the absence of pulpal tissue prevents teeth sensation or further repair.[2] To maintain the functions of pulpal tissue, vital pulp therapy (VPT) is a less invasive alternative treatment to root canal therapy. VPT involves direct pulp capping or partial pulpotomy, in which a biomaterial is used to maintain the vitality of the pulp of a tooth and establish an environment in which the dentin-pulp complex can form.[3] An ideal material for VPT must exhibit favorable sealing ability, nontoxicity to pulp tissue, antibacterial properties, high biocompatibility, stability in tissue fluids, a short setting time, adequate mechanical strength, and favorable handling properties.[4]

Numerous materials, such as zinc oxide eugenol, glass ionomer or resin-modifed glass ionomer, and calcium hydroxide, have been used in VPT[4]; however, none of these materials have achieved the requirements for an ideal material.[5]–[8] Calcium silicate cements, such as Portland cement or mineral trioxide aggregate (MTA), currently exhibit substantial potential for use as biomaterials in VPT.[9], [10] MTA is a bioactive material that features excellent apatite-forming ability; thus, it exhibits a substantial clinical advantage over traditional cements. [11] In addition, numerous studies have shown that MTA exhibits excellent sealing ability, a high pH, radiopacity, biocompatibility, and an ability to stimulate dentin matrix protein expression.[12]–[15] However, the poor handling properties and long setting time of MTA limit its clinical application.[4], [16] Effort is required to overcome the shortcomings of MTA as a biomaterial used in VPT.

In previous studies, we developed a partially stabilized cement (PSC) that exhibits similar properties to MTA.[17] PSC is an innovative biomaterial developed to reform some of the weaknesses of MTA. PSC is a refined calcium silicate cement of which the major chemical constituents are tricalcium silicate ($3CaO \bullet SiO_2$; C3S), dicalcium silicate ($2CaO \bullet SiO_2$; C2S), tricalcium aluminate ($3CaO \bullet Al_2O_3$; C3A), and calcium aluminoferrite ($4CaO \bullet Al_2O_3 \bullet Fe_2O_3$; C4AF), with specific ratios of each component. Among these components, C3S is associated with long-term mechanical strength. However, C3S and C2S exhibit long setting times and low mechanical strength at the early stages of hydrated calcium silicate cement.[18] Nevertheless, C3A exhibits the fastest hydration rate and provides the initial mechanical strength of calcium silicate cement.[19] Therefore, the C3A/C3S ratio may play a crucial role in the early hydration reaction of PSC, producing an accelerating effect on setting time. The sol-gel process is a useful method for preparing ceramics that enables easy control of the compositions of mixtures. Therefore, PSCs with different molar ratios of C3A were fabricated using a one-step sol-gel process featuring a low processing temperature, high chemical homogeneity, uniform phase distribution in a multicomponent system, and high reactivity of the product.[17], [20]

The purposes of this study were to investigate the effect of C3A on PSCs and compare the PSCs with MTA by using X-ray diffraction analysis (XRD) and scanning electron microscopy (SEM) to observe the microstructures and hydration behavior of PSCs in a physiological environment and by evaluating the pH variation, setting time, mechanical properties, and biocompatibility of the cements.

MATERIALS AND METHODS

Material Preparation

PSC powder was prepared using the sol-gel process as previously described.[17] Figure 1 shows schematic diagrams of the preparation. Aluminum *sec*-butoxide (ASB, Al(OBu^s)$_3$), an Al precursor mixed with acetylacetone (acac), was used as the complex ligand for modifying $Al(OBu^s)_3$. The mixture was stirred and reacted for 4 h in a complexing ratio (x) equal to 1. A $Al(OBu^s)_3/(acac)$ complex was formed before conducting further sol-gel process reaction. The complexing ratio (x) represented the molar ratio of acac to $Al(OBu^s)_3$ ($Al(OBu^s)_3/acac$). After the surface of $Al(OBu^s)_3$ was modified, tetraethyl orthosilicate ($Si(OEt)_4$) was added as a Si precursor to the solution, which was subjected to continual stirring. An aqueous solution of $Ca(NO_3)_2$ as a Ca precursor and $Fe(NO_3)_3$ as a Fe precursor was subsequently added to the solution. Ammonia water was added as a catalyst to facilitate reaction between alkoxides. A gel was formed and maintained for 24 h until gelation occurred. The gel was dried at 110°C, and then heated at 1400°C for 2 h and quenched in air. All of the reagents and chemicals used in this study were purchased from Sigma-Aldrich Co (St. Louis, MO, USA). White MTA was obtained from Dentsply, Tulsa Dental Products (Tulsa, OK, USA).

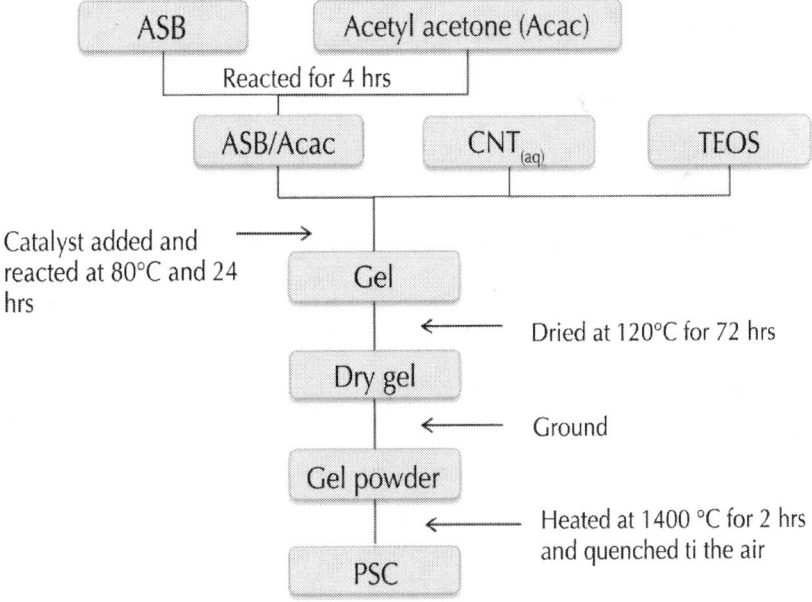

Figure 1: Schematic diagrams of the preparation of PSC. ASB = aluminum sec-butoxide; CNT = calcium nitrate; TOES = tetraethyl orthosilicate; PSC = partial stabilized cement.

Three specimens (PSCs) were prepared in C3S (x)/C3A (y) molar ratios of 9/1, 7/3, and 5/5 (PSC-91, PSC-73, and PSC-55) by using sol-gel processes. De-ionized water (D.I. water) was then added to obtain PSC homogeneous pastes. The liquid-to-powder ratio (L/P) for all specimens was 0.5 mL/g. The mixtures were mixed for 5 min and then placed into a Teflon mold. The specimens were subsequently retrieved from the mold and stored in a sealed container with 100% relative humidity at 37°C to solidify further. The hydration products of all of the specimens were mixed with D.I. water and hydrated with a simulated body fluid (SBF)[21]solution at 4 h, 12 h, 1 day, 3 days, 7 days, 10 days, and 28 days. After incubating for a period of time, the specimens were soaked immediately in an anhydrous ethanol to stop the hydration reaction and enable them to be subjected to material tests.

X-ray Diffraction Analysis

The crystalline phases of all specimens before and after hydration were determined using a Rigaku X-ray powder diffractometer (Rigaku Geigerflex, Japan) with CuK_α radiation ($\lambda = 1.54$ Å) and a Ni filter which was generated at 30 kV and 20 mA. The scanning rate of the specimens was 3°/min, and the scanning range (2θ) was 10° to 60°. The XRD patterns were collected and analyzed according to a model automatched to the standard JCPDS database by using Jade 6.0 software.

Scanning Electron Microscope Observation

The microstructures of the hydration products on the specimen surfaces were examined using a field-emission scanning electron microscope (FE-SEM, Hitachi S-4700) operated at 15 kV. Three specimens (PSCs) with different molar ratios of C3A were prepared for morphological observation conducted using a FE-SEM. After the specimens were air dried for 24 h at room temperature, the surfaces of the hydrated cement specimens were coated with a gold film by using sputtering physical vapor deposition and examined using the FE-SEM.

Vicat Setting Time

The initial and final setting times of the PSCs and MTA were measured using a Vicat needle apparatus. This test was based on International Standard ISO 9597. The Vicat needle was cylindrical and 2.0 mm in diameter. The needle was initially fixed on a 100-g moveable rod and moved in a vertical alignment. Cement was placed in a mold and the needle penetrated 3–5 mm above the bottom of cement paste. The final setting time was determined using the needle (1.13 mm in diameter) loaded on a 300-g moveable rod, which no longer penetrated or indented the surface of the paste. Five cement specimens were measured, and the values are expressed as mean ± standard deviation (mean ± SD).

pH Variation

The pH values of all PSCs and the MTA were measured using the temperature-compensated electrode of a pH meter. Six samples of each cement were subjected to measurement. Each specimen was placed in a tube containing 10 mL of D.I. water and sealed in a container. The pH of the D.I. water in the tube was assessed using a pH meter at 2, 4, 12, 24, 72, 168, and 240 h. After each test, the samples were removed from the container and placed in a new container with the same volume of D.I. water.

Mechanical Strength Measurements

The mechanical strength of all tested cements was evaluated according to the compressive strength. The specimen was placed into a cylindrical Teflon mold (4 mm in diameter × 6 mm in height) and stored in an environment of 100% relative humidity for 24 h at 37°C. Tensile strength data were collected by measuring the diameter and height of the specimens by using a micrometer. The specimens were fractured at a cross-head speed of 1.0 mm/min by using a universal testing machine (Instron 5566, Canton, MA, USA). The maximal load required to fracture each specimen was measured, and the compressive strength, σ, was calculated using the formula $\sigma = \dfrac{P}{A(\pi r^2)}$ where P is the maximum load applied to the specimen in Newtons, and A is the area in millimeters squared. Statistical analysis was conducted using a one-way ANOVA in which pvalues (*) <0.05 indicated significance.

Culture of Human Dental Pulp Cells

Primary human dental pulp cells were used in this study, and approved by the National Taiwan University Hospital (NTUH) Research Ethics Committee (REC) and all patients signed written informed consent, which was obtained from all subjects, dental pulp tissues were obtained from freshly extracted premolars and third molars without caries or pulpal diseases. A tissue explant technique was processed to cultivate dental pulp cells as described previously.[22], [23] Briefly, pulp tissues were minced into small pieces (approximately 1 mm³) and then digested with 3 mg/mL collagenase type I (Sigma, St Louis, MO, USA) and 4 mg/mL dispase (Sigma, St Louis, MO, USA) for 1 h at 37°C. The human dental pulp cells were cultured in Dulbecco's modified Eagle's medium (DMEM), which contained 4 mM L-glutamine, 4500 mg/L of glucose, 1 mM sodium pyruvate, and 1500 mg/L of sodium bicarbonate. The culture medium was supplemented with 10% fetal bovine serum, and the dental pulp cells were incubated in a humidified atmosphere of 5% CO_2 at 37°C. When the dental pulp cells proliferated to 90% confluence in DMEM, the confluent cultures were detached using 0.25% Trypsin-EDTA and subcultured in a flask of DMEM to enable expansion. Cultured human dental pulp cells in passage number 3–10 were used for the following studies.

Cytotoxicity Assay and Cell Viability Assay

The cytotoxicity of all tested cements were measured using the Cyto Tox Non-Radioactive Cytotoxicity Assay detection kit (Promega, Madison, WI, USA). The lactate dehydrogenase (LDH) activity was determined using a spectrophotometric assay. Methods for determining LDH involved combining tetrazolium salts with diaphorase. The chemical reactions of the assay are listed as follows:

$$NAD^+ + lactate \rightarrow pyruvate + NADH$$

$$NADH + tetrazolium\ salt \rightarrow NAD^+ + formazan\ (red\ color)$$

The cell cytotoxicity percentage was calculated by quantifying the amount of LDH in the medium from dead cells and dividing the result by the total amount of LDH in the medium and target cell lysate in the sample. Cell cytotoxicity results are expressed as the percentage of LDH released. The cytotoxicity of the cement was tested in accordance with ISO 10993-5. Briefly, 0.2 g of the sample was soaked in 1 mL of DMEM and incubated in 37°C for 72 h. Dental pulp cells were seeded at a density of 5×10^3 cells per well in a 96-well culture plate for 24 h. After 24 h, the extract solution of the sample containing the medium at various concentrations was then added to the culture plate. The plates were incubated for 24 h and 72 h. The LDH released from the medium was measured using an ELISA reader (optical density at 490 nm).

Dental pulp cells were used in a cell proliferation and viability assay. Cells at 3×10^3 cells/well were cultured in a 96-well culture plate containing 100 µL of DMEM per well for 1 day and 3 days. The cell viability of all tested cements was assessed by using a water-soluble tetrazolium salt-1 (WST-1) cell proliferation assay kit (Roche Diagnostics, Mannheim, Germany). This assay depends on cleavage of WST-1 by mitochondrial dehydrogenase in viable cells. The formazan dye produced by viable cells can be quantified. The number of viable cells was measured colorimetrically by using an ELISA reader at an absorbance (optical density, O.D.) of 440 nm. The results were expressed as the mean O.D. of experimental groups (n = 6) vs. negative control (normal DMEM). The mean O.D. of the control group was set to represent 100% cell viability.

Alizarin Red S Staining

Mineralization of human dental pulp cells was assessed using Alizarin Red S staining (Sigma-Aldrich, St. Louis, MO, USA). Human dental pulp cells (5×10^3 cells/well) were cultured in DMEM containing 10% FBS and treated with the extracts of all tested cements for 21 days for a mineralized nodule assay. The pulp cells were cultured with a culture medium containing extracts of the cements, and the culture medium was replaced every 3 days. After 21 days of treatment, the cells were rinsed twice with phosphate-buffered saline (PBS), fixed with 4% paraformaldehyde for 15 min at room temperature, and stained with a 2% (w/v) Alizarin Red S staining solution at a pH level of 4.2 and a temperature of 37°C. Images of Alizarin Red S staining were viewed and photographed under a light microscope

Statistical Analysis

Data are expressed as the mean ± standard deviation (SD). Statistically significant differences from the control group were determined using a one-way factorial ANOVA. Differences with $p$values (*) <0.05 were considered significant.

RESULTS

X-ray Diffraction Analysis

Figure 2 shows the XRD powder patterns of unhydrated and hydrated cements stored in D.I. water. Unhydrated PSC powder was obtained using sol-gel processes after calcination at 1400°C for 2 h and was characterized using XRD (Figure 2 A). Crystalline phases were characterized using standard data from the JCPDS database. Most of the diffraction peaks were identified as structures of C3S, C2S, C3A, and C4AF, which are the major components of PSC. The peaks at the position $2\theta = 33.2$ and 47.6° corresponded to a C3A structure, and the peak intensity increased as the C3A molar ratio increased. The peaks at $2\theta = 23.2°$, 32.1°–32.7° corresponded to C2S and C3S structures. A similar pattern was observed among the unhydrated MTA samples. In addition, the peak at $2\theta = 37.3°$ and 53.9° corresponded to a CaO structure, and that at $2\theta = 27.4°$ corresponded to a Bi_2O_3 structure.

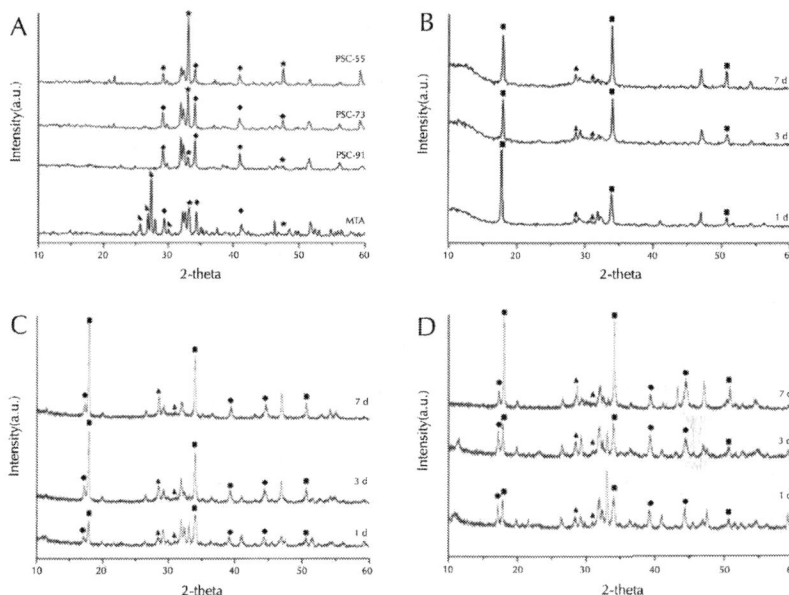

Figure 2: XRD powder patterns of unhydrated cements and hydrated PSCs stored in D.I. water for 1, 3 and 7 day.(A) the unhydrated of PSC-91, PSC-73, and PSC-55 after calcined at 1400°C for 2 h and unhydrated MTA; (B) the hydrated patterns of PSC-91; (C) the hydrated patterns of PSC-73; (D) the hydrated patterns of PSC-55. [★ C3A; ◆ C3S; ▲ Bi_2O_3; ▪ $Ca(OH)_2$; ▲ CSH; • C_3AH_6].

For the hydrated PSCs, three phases of C3S, C2S, and C3A were identified at the same peaks at 2θ. However, the relative intensities of the peaks of these three phases decreased over time after the hydration reaction, as shown in Figure 2 (B)–(D). Portlandite hydrated products and calcium silicate hydrate (CSH) were observed in each experimental group after the samples were stored in D.I. water. Portlandite is a major hydration product of calcium silicate cement. The peaks at the position 2θ = 18° and 34.1° corresponded to $Ca(OH)_2$, and the relative intensity of the peak increased as the degree of hydration increased. CSH (2θ = 28.6°, 29.1°, and 31.6°) was identified in all PSCs because of the hydration of C3S. Figure 2 (C) and Figure 2 (D) show the $Ca_3Al_2(OH)_{12}$ (C_3AH_6) hydration product (2θ = 17.3°, 39.2°, and 44.4°). The relative intensity of the peak of PSC-55 was higher than that of PSC-73 because the amount of C3A in PSC-55 is greater; the intensity of the peak of PSC-91 was unclear (Figure 2 (B)).

Characterization of Hydration Products Soaking In Simulated Body Fluid by Using Scanning Electron Microscopy

The PSC specimens, of which the microstructures are illustrated in Figure 3, were stored in an SBF environment for 1 day and 7 days. Morphological observations indicated that the hydrated PSC-91 soaked in the SBF for 1 day exhibited a cubic microstructure containing mesh-like crystals, as shown in Figure 3 (A). Hydroxyapatite-like crystals, which covered the specimen surface, were the hydration products of PSC-91; no cubic structure was observed. By contrast, the hydrated PSC-73 and PSC-55 that soaked in SBF for 1 day exhibited both a slight amorphous silk-like structure and a crystalline hexagonal structure, as shown in Figures 3 (B) and 3 (C). A silk-like structure is the initial hydration product of calcium silicate with a low Ca/Si ratio and crystallinity, whereas a crystalline hexagonal structure is a product of portlandite, which has a crystallinity that is higher than that of CSH. Figure 3 (E) shows the surface morphology of PSC-91 hydrated in SBF for 7 days; a large portion of the surface was covered with mesh-like HAp crystals. A hydroxyapatite-like structure began to form after only 1 day. After 7 days, aging completely covered the PSC-91 and MTA surface, as shown in Figure 3 (D) and (H). Figure 3 (G) shows that mesh-like crystals embedded the ball-like structure in the matrix of the PSC-55 specimen that was soaked in SBF for 7 days. The hydration products of PSC-55 included more hydroxyapatite-like crystals compared with the specimen that was soaked in SBF for 1 day. A gradual change in the microstructure of PSC-73 was observed using FE-SEM, as shown in Figure 3 (F); additional ball-like crystals and few hydroxyapatite-like crystals covered the matrix.

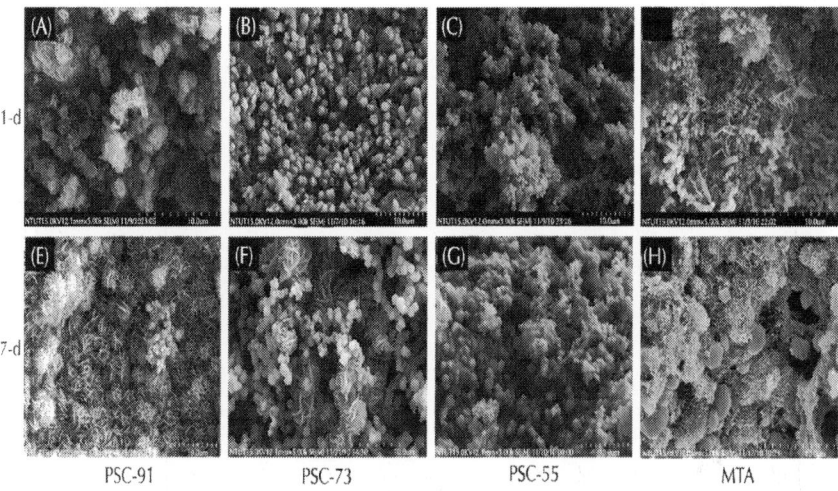

Figure 3: SEM micrograph of the surface of specimens stored in simulated body fluid (SBF) various durations.Soaked in SBF after 1 day: (A) PSC-91 (B) PSC-73 (C) PSC-55 (D) MTA; Soaked in SBF for 7 days: (E) PSC-91 (F) PSC-73 (G) PSC-55 (H) MTA.

Physicochemical Properties of the Cements

The physicochemical properties of all tested cements are shown in Figure 4. The change in pH as a function of time for all materials is shown in Figure 4 (A). The initial pH value of all tested cements after mixing was approximately 7.5 and rose to 12.4–12.7. The initial pH value of PSC-91, PSC-73, and PSC-55 at 4 h was higher than that of the MTA, and the alkalinity of these PSCs tended to increase over time. All cements exhibited a high alkaline pH. The highest value was that of PSC-91, and the mean was 12.7 at 240 h. The pH values of PSC-91, PSC-73, and PSC-55 did not differ significantly from those of MTA after 72 h.

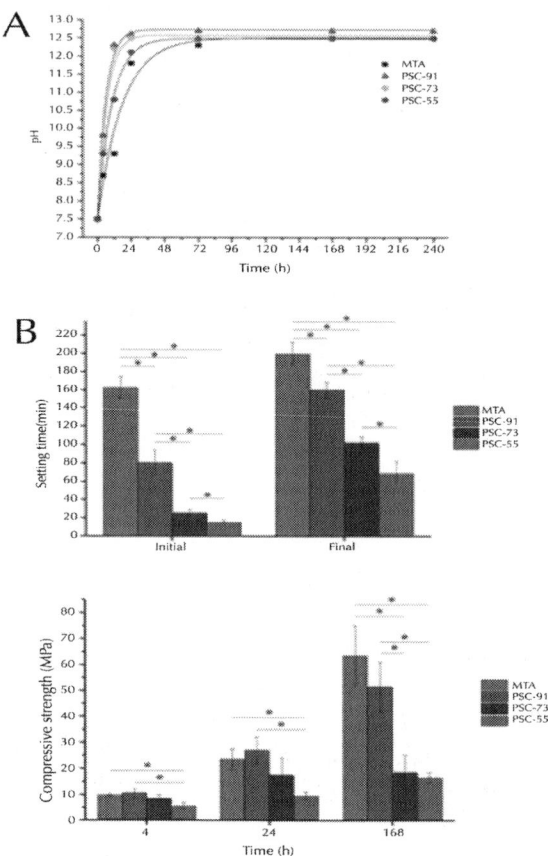

Figure 4: The pH value and physical properties of all tested cements. (A) pH values variation of all tested cements at various time intervals. (B) The initial and final setting times of PSC-91, PSC-73, PSC-55 and MTA. (C) Compressive strength of PSC with different C3S and CSA content and MTA at 4-h, 24-h and 168-h.

The initial and final setting times of all of the materials are shown in Figure 4 (B). The initial setting times for PSC-91, PSC-73, and PSC-55 were between 15.5 and 80.5 min, and the final setting times ranged between 68.5 and 160 min. C3A exhibited an accelerated setting effect during hydration of the PSCs compared with MTA. When the molar ratios of C3A in the PSC were increased from 10% to 50%, the initial and final setting times of the PSCs decreased dramatically compared with those of MTA.

Figure 4 (C) shows the compressive strength of all of the tested cements. PSC-91 and PSC-73 exhibited similar compressive strength in the early stage in 4 h. The compressive strength of PSC-55 was slightly lower than that of the other experimental samples at 4 h. After being set in D.I. water for 24 h, PSC-91 exhibited higher compressive strength (27.1 MPa) than PSC-73 and PSC-55 did. PSC-91 exhibited the highest compressive strength among the PSCs, 51.5 MPa, after it was set for 168 h. Reducing the amount of C3A increased the mechanical strength of the PSCs. The compressive strength of PSC-73 and PSC-55 did not differ significantly after the specimens were set for 168 h. The results revealed that the content of C3A in the PSCs exerted an obvious effect on the compressive strength of the cement.

Biocompatibility Assay

The cytotoxicity of all tested cements was determined by conducting an LDH assay. The cytotoxicity of the control group and all of the tested biomaterials increased over time, as shown in Figure 5 (A). PSC-91 and PSC-73 exhibited no statistically significant difference from the MTA at 1 day and 3 days. However, the cytotoxicity of PSC-55 was significantly higher than that of the other experimental samples at 1 day and 3 days. The results of the LDH assay indicated that the PSCs exhibited low levels of cell cytotoxicity.

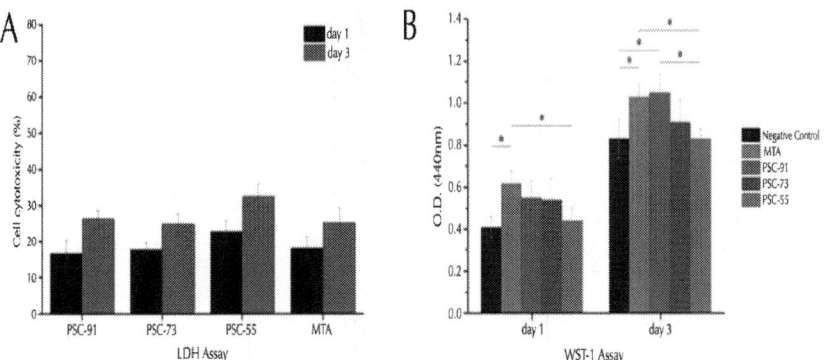

Figure 5: (A) Cytotoxicity assessment of PSC-91, PSC-73, PSC-55 and mineral trioxide aggregate (MTA) by LDH assay according to ISO-10993 protocol standard. All tested cements on dental pulp cells were

evaluated by LDH assay on 1day and 3 days. Each bar illustrated average absorbance (A490 nm) ± SD. No significant differences between PSC-91, PSC-73 and MTA (P>0.05); (B) Cell viability evaluation by WST-1 assay. Each bar illustrated average absorbance (A440 nm) ± SD.

Cell viability was evaluated according to the mitochondrial function (WST-1 assay) of the cells. The O.D. value was directly proportional to cell number. As shown in Figure 5 (B), the O.D. value of MTA group was significant higher than that of the negative control group at day 1 and 3. The O.D. value of PSC-91 was significantly increased compared with negative control group at day 3. The cell numbers of PSC-55 at 1 day and 3 days exhibited a statistically significant difference (p<0.05) from those of MTA. Increasing the molar ratios of C3A in the PSCs decrease the activity of mitochondria (cell viability)

Mineralized Nodule Formation

The effect of all of the tested cements on the formation of calcification nodules in human dental pulp cells that were incubated for 21 days and examined using Alizarin Red S staining is shown in Figure 6. PSC-91 and MTA exhibited a significant increase in the area of calcified nodules compared with PSC-73 and PSC-55, whereas no clear mineralization was observed in the control cells.

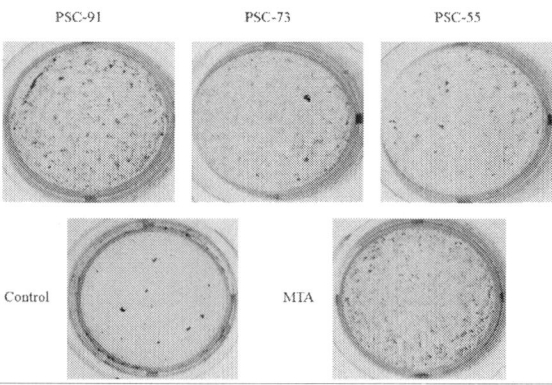

Figure 6: The evaluation of biomineralization in human dental pulp cells by Alizarin Red S staining.

DISCUSSION

In this study, three PSCs containing C3A in different ratios were fabricated using sol-gel processes. The sol-gel process is a useful method for preparing ceramics and glass [24] and features high chemical homogeneity, uniform phase distribution in multicomponent systems, high reactivity of the product, and low synthesis temperatures. According to the XRD results (Figure 2 (A)), the major components of the synthesized unhydrated PSCs were C3S, C2S, and C3A; this composition is similar to those of MTA and portlandite. The XRD peak intensity of C3A increased as the molar ratio of C3A in PSC increased, indicating that a proportion of raw materials reacted homogenously during the sol-gel process.

C3S and C2S are two major components of PSC. Their hydration reaction can be expressed using Formulas (1)–(3). The hydration products are CSH and calcium hydroxide (CH). The C3A is substantially influenced by the early hydration behavior of PSC. The hydration of C3A is rapid and its hydration reactions can be expressed using Formula (4). [25] The hydration process in this reaction involves several hydrates; ($Ca_3Al_2(OH)_{12}$; C_3AH_6) is one of the hydrates and is the most stable at high temperatures.

The hydration reaction of each component of PSC is expressed as follows:

$$2C3S + 6H \rightarrow C_3S_2H_3 + 3CH$$

(1)

$$2C2S + 4H \rightarrow C_2S_2H_2 + 2CH$$

(2)

$$C3S + H \underset{K_2}{\overset{K_1}{\rightleftharpoons}} C3SH_n \underset{K_4}{\overset{K_3}{\rightleftharpoons}} C-S-H + CH$$

(3)

$$C3A + H \rightarrow gel \rightarrow irregular\ flakes \rightarrow hexagonol\ flakes \rightarrow$$

$$C_3AH_6\ single\ crystals \rightarrow C_3AH_6\ aggregates \tag{4}$$

where [A = Al_2O_3; C = CaO; CH = $Ca(OH)_2$; H = H_2O; S = SiO_2; C-S-H is amorphous hydrogen having variable composition in terms of Ca/Si ratio and H_2O/SiO_2 ratios].

The hydrolysis of calcium silicates produces calcium hydroxide and creates a less basic calcium silicate hydrate. [26] Calcium hydroxide precipitated during the hydration of PSC. The presence of calcium hydroxide causes the hydrated PSC to be highly alkaline (pH 12.5). All of the cements evaluated in this study increased the pH values (7.5 at 12.7) through the release of hydroxyl ions. The more C3A in PSC, the slower pH value rising in the immersed solution. The curve of pH versus immersion time; where slope of the initial stage was different depends on the C3A molar ratio in PSCs. Figure 4 (A) shows the rising curve: PSC-91>PSC-73>PSC-55>MTA. MTA contains gypsum, which hindered the hydration reaction, causing the pH value to rise at slowest rate. The highly alkaline environment has an antibacterial effect and promotes cell remineralization.[27]

One of the most clinically relevant factors is the setting time of biomaterials, which is affected by numerous factors, such as the L/P ratio, particle size of the cement, chemical content of the cement, and cement additives. A long setting time may prevent the material from being held at the operation site, causing a lack of mechanical strength required for initial support. [28] The initial setting time of MTA is approximately 165 min. Because of this long initial setting time, a two-step procedure must be used when applying MTA in VPT. C3A is the most reactive component of Portland cement, and increasing the molar ratio C3A in PSCs may reduce the setting time and enhance the initial compressive strength. PSC-55 has short initial (15.5 min) and final (68.5 min) setting times (Figure 4 (B)) because the ratio of C3A to C3S is high; however, the compressive strength of PSC-55 and PSC-73 was lower than that of the other experimental groups during the 7-day hydration reaction, as shown in Figure 4 (C). By contrast, C3S and C2S play a crucial role in cement mechanical strength, and C2S hydration occurs more slowly than C3S hydration does, contributing to the long-term strength of the cement.

Calcium silicate materials exhibit dissolution and precipitation behaviors during the hydration reaction. The calcium and hydroxyl ions released from materials reacts with phosphate to form a hydroxyapatite structure. [29] Hydroxyapatite plays a crucial role in tissue regeneration and maintaining function because of its bioactive surface.

According to the SEM observation, the hydroxyapatite layer precipitated on the surface of all of the PSCs after they were exposed to the SBF environment for 7 days, as shown in Figure 3. The mesh-like apatite layer was observed in PSC-91 and MTA at 1 day and 7 days. However, this layer was not observed in PSC-73 and PSC-55 until the samples were immersed in SBF for 7 days. In addition, the hydration products were examined using XRD, and the patterns indicated that Ca-P hydrated products and calcium carbonate ($CaCO_3$) were present in each experimental group after the samples were soaked in SBF. In addition, the characteristic peaks of hydroxyapatite crystals were present in the PSC-91 and MTA groups, suggesting that hydroxyapatite precipitated on the surfaces of all of the specimens after they were soaked in SBF for 7 days. Both the SEM and XRD results indicated that the C3A and C3S in PSCs facilitated the precipitation of a hydroxyapatite-like active layer, suggesting that the PSCs feature favorable in vitro bioactivity. The time required for C3A and C3S to induce a hydroxyapatite-like structure increases when the C3A molar ratio is over 30% in PSC. Small round particles were observed on the surface of the PSCs after they were immersed in SBF for various durations. XRD analysis revealed that the hydration products were calcium-deficient carbonated apatite or had a calcium phosphate structure.

The biocompatibility of dental materials can be evaluated using numerous mammalian cell culture methods. Extracts of all of the tested cements were used in this study. According to the ISO 10993-5 standard, all of the tested cements were evaluated using in vitro tests of cytotoxicity, and the response of cells to the extracts were assessed. Numerous studies have reported that MTA features high biocompatibility. [30]–[32] In this study, MTA was used as a comparative biomaterial because PSC is chemically similar to Portland cement, which exhibits a biological response similar to that of MTA. [33] Figure 6 (A) indicates that PSC-55 contained a substantial amount of C3A, exhibiting a statistically significant difference in the LDH assay compared with the other groups at 1 day and 3 days of incubation. Liu et al. determined that adding 0%–15% C3A into C3A and C3S mixtures yields no significant cytotoxicity within the extract concentration range between 3.125 mg/mL and 100 mg/mL. [34] The extract concentration in this study was 100 mg/mL and, according to the results of the cell viability assay, all of the PSCs exhibited no cytotoxicity and did not influence cell function for a long period of time. Finally, LDH and WST-1 assay kits were used in this study, and the results indicated that all of the PSCs and MTA are biocompatible. Mineralized nodules were characterized by using Alizarin Red S staining. After 21 days of induction, we observed that mineralized nodules of human dental pulp cells formed. All of the tested cements induced mineralization of the pulp cells. In addition, the stain of mineralized nodules of PSC-91 and MTA was more intense under light microscopy. These results indicated that PSC-91 can facilitate the formation of mineralized nodules of human dental pulp cells.

In summary, increasing the C3A content from 30% to 50% significantly improved the setting time of the PSCs; however, the mechanical strength and biological properties of the PSCs deteriorated. According to the results of this study, PSC-91 exhibits an appropriate setting time and high mechanical strength as well as a cellular response similar to that of MTA, a commercial product applied in VPT. Based on its physicochemical properties, in vitro biocompatibility, and bioactivity, PSC-91 can be applied in VPT.

CONCLUSIONS

In this study, three PSCs containing C3A in different molar ratios were fabricated using sol-gel processes. Their physicochemical properties, bioactivity, and biocompatibility were characterized using XRD, SEM, pH-metry, an Instron machine, a Vicat needle apparatus, Alizarin Red S staining, and a cytotoxicity assay kit. The PSC containing 10% C3A (PSC-91) exhibited optimal compressive strength and a setting time shorter than that of MTA. The hydration properties of C3A in PSCs played a critical role at the early stages, and the results indicated that PSC-91 facilitated apatite formation when the cement was soaked in SBF. The results of the Alizarin Red S staining and biocompatibility assay indicated that PSC-91 exerts no effect on cell viability or cytotoxicity and facilitates the formation of mineralized nodules of human dental pulp cells. The physicochemical properties and biocompatibility of

PSC-91 indicated that this material has considerable potential for use as a biomaterial in VPT. However, additional studies are required to explore the physical properties, antimicrobial properties, and in vivo effectiveness of this material.

AUTHOR CONTRIBUTIONS

Conceived and designed the experiments: KCC FHL CPL. Performed the experiments: KCC CCC YCH MHC. Analyzed the data: CCC YCH. Contributed reagents/materials/analysis tools: YCH MHC. Wrote the paper: KCC CCC CPL.

REFERENCE

1. Schmalz G, Smith AJ (2014) Pulp development, repair, and regeneration: challenges of the transition from traditional dentistry to biologically based therapies. J Endod 40: S2–5.

2. Zhang W, Yelick PC (2010) Vital pulp therapy-current progress of dental pulp regeneration and revascularization. Int J Dent 2010: 856087.

3. Witherspoon DE (2008) Vital pulp therapy with new materials: new directions and treatment perspectives–permanent teeth. J Endod 34: S25–28.

4. Hilton TJ (2009) Keys to Clinical Success with Pulp Capping: A Review of the Literature. Operative Dentistry 34: 615–625.

5. Ho YC, Huang FM, Chang YC (2006) Mechanisms of cytotoxicity of eugenol in human osteoblastic cells in vitro. Int Endod J 39: 389–393.

6. Costa CA, Giro EM, do Nascimento AB, Teixeira HM, Hebling J (2003) Shortterm evaluation of the pulpo-dentin complex response to a resin-modified glassionomer cement and a bonding agent applied in deep cavities. Dent Mater 19: 739–746.

7. Kitasako Y, Ikeda M, Tagami J (2008) Pulpal responses to bacterial contamination following dentin bridging beneath hard-setting calcium hydroxide and self-etching adhesive resin system. Dent Traumatol 24: 201–206.

8. Prosser HJ, Groffman DM, Wilson AD (1982) The effect of composition on the erosion properties of calcium hydroxide cements. J Dent Res 61: 1431–1435.

9. Bogen G, Kim JS, Bakland LK (2008) Direct pulp capping with mineral trioxide aggregate: an observational study. J Am Dent Assoc 139: 305–315; quiz 305– 315.

10. Karabucak B, Li D, Lim J, Iqbal M (2005) Vital pulp therapy with mineral trioxide aggregate. Dent Traumatol 21: 240–243.

11. Gandolfi MG, Taddei P, Tinti A, Prati C (2010) Apatite-forming ability (bioactivity) of ProRoot MTA. Int Endod J 43: 917–929.

12. Islam I, Chng HK, Yap AU (2006) Comparison of the physical and mechanical properties of MTA and portland cement. J Endod 32: 193–197.

13. Fridland M, Rosado R (2005) MTA solubility: a long term study. J Endod 31: 376–379.

14. Torabinejad M, Hong CU, McDonald F, Pitt Ford TR (1995) Physical and chemical properties of a new root-end filling material. J Endod 21: 349–353.

15. Tomson PL, Grover LM, Lumley PJ, Sloan AJ, Smith AJ, et al. (2007) Dissolution of bio-active dentine matrix components by mineral trioxide aggregate. J Dent 35: 636–642.

16. Malhotra N, Agarwal A, Mala K (2013) Mineral trioxide aggregate: a review of physical properties. Compend Contin Educ Dent 34: e25–32.

17. Wang WH, Lee YL, Lin CP, Lin FH (2008) Synthesis of partial-stabilized cement (PSC) via sol-gel process. J Biomed Mater Res A 85: 964–971.

18. Grech L, Mallia B, Camilleri J (2013) Investigation of the physical properties of tricalcium silicate cement-based root-end filling materials. Dent Mater 29: e20– 28.

19. Oh SH, Choi SY, Choi SH, Lee YK, Kim KN (2004) The influence of lithium fluoride on in vitro biocompatibility and bioactivity of calcium aluminatepMMA composite cement. J Mater Sci Mater Med 15: 25–33.

20. Meiszterics A, Rosta L, Peterlik H, Rohonczy J, Kubuki S, et al. (2010) Structural characterization of gel-derived calcium silicate systems. J Phys Chem A 114: 10403–10411.

21. Kokubo T, Takadama H (2006) How useful is SBF in predicting in vivo bone bioactivity? Biomaterials 27: 2907–2915.

22. Chang MC, Chen YJ, Tai TF, Tai MR, Li MY, et al. (2006) Cytokine-induced prostaglandin E2 production and cyclooxygenase-2 expression in dental pulp cells: downstream calcium signalling via activation of prostaglandin EP receptor. Int Endod J 39: 819–826.

23. Chang HH, Chang MC, Huang GF, Wang YL, Chan CP, et al. (2012) Effect of triethylene glycol dimethacrylate on the cytotoxicity, cyclooxygenase-2 expression and prostanoids production in human dental pulp cells. Int Endod J 45: 848–858.

24. Gupta R, Kumar A (2008) Bioactive materials for biomedical applications using sol-gel technology. Biomed Mater 3: 034005.

25. Jupe AC, Turrillas X, Barnes P, Colston SL, Hall C, et al. (1996) Fast in situ xray-diffraction studies of chemical reactions: A synchrotron view of the hydration of tricalcium aluminate. Phys Rev B Condens Matter 53: R14697–r14700.

26. Camilleri J (2007) Hydration mechanisms of mineral trioxide aggregate. Int Endod J 40: 462–470.

27. Morgental RD, Vier-Pelisser FV, Oliveira SD, Antunes FC, Cogo DM, et al. (2011) Antibacterial activity of two MTA-based root canal sealers. Int Endod J 44: 1128–1133.

28. Ishikawa K, Miyamoto Y, Takechi M, Toh T, Kon M, et al. (1997) Non-decay type fast-setting calcium phosphate cement: hydroxyapatite putty containing an increased amount of sodium alginate. J Biomed Mater Res 36: 393–399.

29. Samachson J (1969) Basic requirements for calcification. Nature 221: 1247–1248.

30. Torabinejad M, Parirokh M (2010) Mineral trioxide aggregate: a comprehensive literature review–part II: leakage and biocompatibility investigations. J Endod 36: 190–202.

31. Keiser K, Johnson CC, Tipton DA (2000) Cytotoxicity of mineral trioxide aggregate using human periodontal ligament fibroblasts. J Endod 26: 288–291.

32. Lee SK, Lee SK, Lee SI, Park JH, Jang JH, et al. (2010) Effect of calcium phosphate cements on growth and odontoblastic differentiation in human dental pulp cells. J Endod 36: 1537–1542.

33.   Dreger LA, Felippe WT, Reyes-Carmona JF, Felippe GS, Bortoluzzi EA, et al. (2012) Mineral trioxide aggregate and Portland cement promote biomineralization in vivo. J Endod 38: 324–329.

34.   Liu WN, Chang J, Zhu YQ, Zhang M (2011) Effect of tricalcium aluminate on the properties of tricalcium silicate-tricalcium aluminate mixtures: setting time, mechanical strength and biocompatibility. Int Endod J 44: 41–50

A Novel Injectable Calcium Phosphate Cement-Bioactive Glass Composite for Bone Regeneration

Long Yu[1], Yang Li[1], Kang Zhao[2], Yufei Tang[2],
Zhe Cheng[2], Jun Chen[1], Yuan Zang[1], Jianwei Wu[1],
Liang Kong[3], Shuai Liu[1], Wei Lei[1],
and Zixiang Wu[1]

[1]Institute of Orthopedics, Xijing Hospital, The Fourth Military Medical University, Xi'an, Shaanxi Province, People's Republic of China

[2]School of Materials and Engineering, Xi'an University of Technology, Xi'an, Shaanxi Province, People's Republic of China

[3]Department of Oral and Maxillofacial Surgery, School of Stomatology, The Fourth Military Medical University, Xi'an, Shaanxi Province, People's Republic of China

ABSTRACT

Background

Calcium phosphate cement (CPC) can be molded or injected to form a scaffold in situ, which intimately conforms to complex bone defects. Bioactive glass (BG) is known for its unique ability to bond to living bone and promote bone growth. However, it was not until recently that literature was available regarding CPC-BG applied as an injectable graft. In this paper, we reported a novel injectable CPC-BG composite with improved properties caused by the incorporation of BG into CPC.

Materials and Methods

The novel injectable bioactive cement was evaluated to determine its composition, microstructure, setting time, injectability, compressive strength and behavior in a simulated body fluid (SBF). The in vitro cellular responses of osteoblasts and in vivo tissue responses after the implantation of CPC-BG in femoral condyle defects of rabbits were also investigated.

Results

CPC-BG possessed a retarded setting time and markedly better injectability and mechanical properties than CPC. Moreover, a new Ca-deficient apatite layer was deposited on the composite surface after immersing immersion in SBF for 7 days. CPC-BG samples showed significantly improved degradability and bioactivity compared to CPC in simulated body fluid (SBF). In addition, the degrees of cell attachment, proliferation and differentiation on CPC-BG were higher than those on CPC. Macroscopic evaluation, histological evaluation, and micro-computed tomography (micro-CT) analysis showed that CPC-BG enhanced the efficiency of new bone formation in comparison with CPC.

Conclusions

A novel CPC-BG composite has been synthesized with improved properties exhibiting promising prospects for bone regeneration.

INTRODUCTION

Calcium phosphate biomaterials, such as hydroxyapatite (HA) ceramic, calcium phosphate ceramics and calcium phosphate cements (CPC), have been widely used as bone substitute materials in clinical applications due to their good biocompatibility and osteoconduction [1]. However, it is difficult to fill irregularly shaped bone defects with sintered bioactive ceramics and these materials have obvious limitations for use in minimally invasive surgery [2]. CPC can be molded or injected to form a scaffold in situ that intimately conforms to the shape of complex bone defects [3]. In 1986, a typical CPC composed of a powdered mixture of tetracalcium phosphate (TECP) [$Ca_4 (PO_4)_2O$] and dicalcium phosphate anhydrous (DCPA) ($CaHPO_4$) was first reported by Brown and Chow [4]. This CPC powder could be mixed with aqueous liquid to form a paste that would set in situ and form HA ($Ca_{10} (PO_4)_6 (OH)_2$) as a final product, as the main constituent part of the mineral phase of bone [5]. Due to its high biocompatibility, osteoconduction and bone replacement capability, CPC was approved in 1996 by the Food and Drug Administration (FDA) to repair craniofacial defects [6]. Since then, several other calcium phosphate cements and injectable cements have been developed [7], [8]. However, under adverse critical conditions, such as poorly vascularized sites and elderly patients with metabolic disorders, the osteoconductive and degradation properties of CPC are not sufficient to achieve complete bone regeneration. Hence, it is necessary to enrich CPC with osteopromotive or osteoinductive factors to improve its biological performance [9].

As a surface-active bone substitute, bioactive glass (BG) has recently attracted attention due to its good biocompatibility both in bone and in soft tissues [10]. Bioactive glasses e.g. 45S5 bioactive glass bond strongly to bone and promote bone growth [11] by forming a hydroxycarbonate apatite (HCA) layer and releasing Ca, P and Si ions [10]. These ions are supposed to stimulate osteogenesis [12]. 45S5 bioactive glass promoted the attachment, proliferation, differentiation and mineralization of osteoblast-like cells [13]; up-regulated seven families of osteoblast genes in osteoprogenitor cells; increased osteoblast phenotype expression of osteoprogenitor cells; [14] and induced differentiation of bone marrow stromal cells into mature osteoblasts [15], suggesting that this bioactive glass creates both solution-mediated and surface-mediated effects on bone cell activity. However, previous studies have mainly focused on the preparation of BG in the form of a scaffold. Relevant literature only recently became available regarding CPC-BG as applied in minimally invasive injectable grafts [16].

The purpose of the present study was to synthesize a type of self-setting bioactive cement by the incorporation of BG into CPC. The composition, morphology/microstructure, setting time, injectability, compressive strength, surface reaction layer formation and degradation of CPC-BG were investigated, and the cell and tissue responses to CPC-BG were also investigated both in vitro and in vivo.

MATERIALS AND METHODS

Preparation and Characterization of the CPC-BG

CPC consisted of a powder and a liquid phase. The CPC powder was composed of tetracalcium phosphate (TECP) and dicalcium phosphate anhydrous (DCPA) in an equal molar ratio, and the preparation method was as previously described [17]. Briefly, TECP was synthesized by a solid-to-solid reaction between calcium phosphate and calcium carbonate at a temperature of 1500°C for 8 h. Dicalcium phosphate dehydrate (DCPD, $CaHPO_4 \cdot 2H_2O$) was prepared from ammonium hydrogen phosphate (($NH_4)_2HPO_4$) and calcium nitrate ($Ca(NO_3)_2$) in an acidic environment. DCPA was obtained by removing the crystallization water in DCPD at 120°C. The TECP and DCPA powders were then mixed in a micromill to form the CPC powder. All the chemicals used were purchased from Sinopham Chemical Reagent Co. Ltd.

Bioglass 45S5 (Wt %: 45% SiO_2, 24.5% Na_2O, 24.5% CaO and 6% P_2O_5) were provided by NovaBone® (LLC, Alachua, USA). The NovaBone® product were ground in a ball mill and sieved to obtain

45S5 particles with sizes ranging from approximately 5–10 µm, with a median of 6 µm.

The CPC-BG powder was prepared by adding BG 45S5 powder (10 and 20 wt%) into the CPC powder. The CPC-BG composite powders were mixed with potassium phosphate buffers (pH 7.0) for 1 min at the given P/L ratio (2.0 g/mL) with a spatula to form homogeneous paste. The paste was then placed into a plastic cylindrical mold with a diameter of 5 mm and a height of 10 mm for mechanical testing, a circular mold with a diameter of 10 mm and a thickness of 3 mm for measurement of in vitro bioactivity, degradability and cell studies. Each specimen was set in a 100% relative humidity box at 37°C for 24 h, at which time the hardened CPC-BG composite was obtained. X-ray diffraction (XRD, Dandong fangyuan Co., China) with CuK radiation in a continuous scan mode was adopted to characterize the phase composition of the phase composition of the specimens. The 2 range was from 10° to 90° at a scanning speed of 2.4°/min. The cross-section of the specimens was examined with a scanning electron microscopy (SEM, Hitachi S-4800, Japan) equipped with an energy dispersive spectrometer (EDS, Falcon, USA).

Setting Time, Injectability and Compressive Strength of CPC-BG

CPC-BG paste was placed into a plastic cylindrical mold with a diameter of 5 mm and a height of 10 mm, and it was then allowed to set in a 100% relative humidity box at 37°C.

The setting time was taken as the time at which the paste hardened to such an extent that a needle (300 g, = 1 mm) would not penetrate deeper than 1 mm into the sample. This criterion to determine the setting time is based on the American Society for Testing and Materials C 187-98 standard test method for normal consistency of hydraulic cement, which is also called the Vicat method, to determine the setting time. Each specimen was performed in triplicate and the average value was calculated.

The injectability of CPC-BG composite paste was evaluated by extruding 2.0 g of as-prepared paste through a 2.5 mL disposable syringe with an opening nozzle with the diameter of 2.0 mm by hand, according to a modified method described previously [18], suggesting that injection by hand possessed even slightly lower standard deviations than injection by machine with preset load. After setting at 37°C in a 100% relative humidity box for pre-selected time, the paste was extruded from the syringe until it was unable to be injected. The weight of the paste injected through the syringe was measured. The injectability was calculated as: $I = m_{injected}/m_{initial} \times 100\%$, where I is the injectability, $m_{injected}$ and $m_{initial}$ are the weight of the paste injected through the syringe and the paste initially contained in the syringe. All values were the average of three tests performed for each group and presented as mean ± standard deviation (mean ± SD).

After hardened at 37°C in a 100% relative humidity box for 1 and 7 days, the compressive strength of CPC-BG composite specimens was measured at a loading rate of 1 mm/min with a universal testing machine (MTS-858, MTS System Inc, USA). The compressive strength was calculated as following: $S = Fmax/A$, where Fmax is the maximum load on the load-deformation curve and A is the cross-sectional area of each specimen. The measurement was performed three times and the results were expressed as mean ± SD.

Bioactivity and Degradation in Simulated Body Fluid (SBF)

Simulated body fluid (SBF), which has ion concentrations and a pH value similar to those of human blood plasma, was prepared in accordance with the procedure described [19]. After setting for 24 h, the paste specimens (10 mm diameter and 3 mm thickness) were soaked in an SBF solution at 37°C for 7 and 14 days with a weight-to-volume ratio of 0.2 g/ml and the solution was refreshed every day. For the evaluation of in vitro bioactivity, the samples were removed after incubation for specified time periods, rinsed in deionized water and dried at room temperature until a constant weight was attained. The specimens were characterized with EDS, and the surface morphologies of the specimens were observed with SEM.

For the measurement of in vitro degradation, the 7-day-set paste specimens were immersed into SBF solution at 37°C for 28 days with a weight-to-volume ratio of 0.2 g/ml and the solution was refreshed every day. For this evaluation, the samples were removed after incubation for specified time periods, rinsed in deionized water and dried at 60°C for

24 h and weighed. In vitro degradation was measured as $D = [(W_0 - Wt)/W_0] \times 100\%$, where D is the degradation rate and W_0 and Wt are the dry weight of the initial specimen and the degraded specimen, respectively. All values presented are the average of five tests performed for each sample.

Cell Attachment, Proliferation, Morphology and Differentiation

With protocols approved by the Institutional Animal Care Committee of Xijing Hospital (Permit Number: 08–269), osteoblasts were obtained from calvariae of 1-to 2-day-old Sprague-Dawley rats with an enzyme-digestion technique as previously described [20]. Briefly, under general anesthesia, calvariae were dissected aseptically, rinsed with PBS several times, stripped of the periosteum and adherent tissue and then minced. The minced fragments were incubated with 4 ml of 0.25% trypsin (Sigma, USA) for 20 min with gentle shaking at room temperature. Then, the trypsin was removed and fetal bovine serum (FBS, HyClone, USA) was added to terminate digestion. After digestion, the fragments were uniformly attached to each culture flask bottom and were cultured in a humidified atmosphere of 5% CO_2 at 37°C. After 24 h, Dulbecco's modified Eagle's medium (DMEM, Sigma, USA) supplemented with 1% penicillin/streptomycin antibiotics (Sigma, USA) and 10% FBS was added to the flasks. The medium was changed every 2 days. Cell subcultures at passage 2 were used in the following studies.

To investigate cell attachment and proliferation, the CPC and CPC-BG composite specimens (10 mm diameter and 3 mm thickness) were sterilized by ^{60}Co irradiation for 12 h. Prior to cell seeding, CPC and CPC-BG composite specimens were pre-wetted with basal tissue culture medium (DMEM supplemented with 1% penicillin/streptomycin antibiotics and 10% FBS) after setting for 24 h. Osteoblasts with a density of 5×10^4 cells/well were seeded onto the specimens. Cell attachment study was determined with a 3-(4, 5-dimethylthiazol-2-yl)-2, 5-diphenyl tetrazolium bromide (MTT) (Sigma, USA) assay. The specimen-cell constructs were incubated in a humidified atmosphere of 5% CO_2 at 37°C for 4, 8 h, rinsed in 0.15 M PBS, immersed in a mixture of serum-free cell culture medium and MTT reagent (5:1), followed and incubated in a humidified atmosphere of 5% CO_2

at 37°C for 4 h. The supernatant from each well was then carefully removed and dimethyl sulfoxide was added to ensure the solubilization of crystals. Next, 150 µl of the reaction solution with the cells was carefully transferred to 96-well plates, and the optical density (OD) values at 492 nm were measured. Six specimens in each group were tested for each incubation period; each test was carried out in triplicate per specimen. The results were presented as means ± SD.

Cell proliferation was evaluated by seeding cells at a density of 5×10^4 cells per sample followed by incubation for 1, 4 and 7 days; the medium was replaced every second day. Adhesion and cell viability on the substrates were assessed quantitatively using the MTT assay. Six specimens in each group were tested for each incubation period; each test was carried out in triplicate per specimen. The results were presented as means ± SD.

Cell attachment and morphology were confirmed by direct visualization of specimen-cell constructs under SEM. The cells were attached to the specimens for 1 and 4 days in a humidified atmosphere of 5% CO_2 at 37°C. At a pre-selected time point, the specimen-cell constructs were removed, rinsed with PBS twice and fixed with 2.5% glutaraldehyde in 0.1 M PBS for 30 min. The fixed constructs were washed with PBS three times, dehydrated in graded ethanol, vacuum-dried at 37°C overnight and sputter-coated with gold-palladium prior to SEM observation.

To evaluate ALP activity, the medium was aspirated and specimens were moved to new 24-well plates after 4 and 7 days of incubation. Approximately 500 µl of cell lysis buffer containing Triton X-100 was added to each well at room temperature to lyse the cells. The cell lysate was placed in a 1.5 mL centrifuge tube, centrifuged and then frozen to −20°C. At 4 and 7 days, the frozen samples were thawed at room temperature for 5 min to measure ALP activity following the manufacturer's instructions (Sigma, USA). The absorbance of ALP was quantified with a plate reader at 405 nm. The total protein content was determined by the bicinchoninic acid method using a Pierce protein assay kit (Pierce Biotechnology Inc., USA) according to the manufacturer's instructions. The ALP activities were normalized to the total protein content. Each experiment was carried out in triplicate, and the results were presented as means ±SD.

Implantation in vivo

Eighteen healthy female New Zealand white rabbits aged 4–5 months and weighing 2.5–3 kg were randomly divided into three groups of 6 animals for each type of implant. The experiment was carried out in strict accordance with the recommendations in the Guide for the Care and Use of Laboratory Animals for the National Institutes of Health. The animal protocol was approved by the Institutional Animal Care Committee of Xijing Hospital (Permit Number: 08–269). Under general anesthesia and sterile conditions, cylindrical specimens (6 mm diameter and 10 mm height) were implanted into each femoral condyle. Then, the wounds were sutured and penicillin (240, 000 UI) was injected into the rabbits for 3 days. Three animals of each group were sacrificed with an overdose abdominal injection of pentobarbital sodium at 4 and 12 weeks after implantation. The bone specimens were harvested immediately after sacrifice and fixed in 10% neutral buffered formalin. The macroscopic appearance of the defects was evaluated to assess the degree of specimen incorporation and tissue reactions adjacent to the specimens. For the micro-computed tomography (micro-CT) analysis, the bone specimens were imaged with three-dimensional microfocus computed tomography (micro-CT, eXplore Locus SP, GE, USA) at a voltage of 80 kVp and an electric current of 80 mA. To evaluate the in vivo resorption of the implanted materials, the residual material volume fraction (RMVF) was calculated as RMVF = VR/VT where VR is the volume of residual material and VT is the total volume of material. The new bone volume was quantified as the bone volume fraction (BVF) with the formula BVF = VB/VT where VB is the newly formed bone volume and VT is the total material volume. For histological evaluation, the bone specimens were embedded in methacrylate resin after micro-CT scanning. Tissue blocks were sectioned to 5 μm thickness, and at least three slices of histology sections were randomly obtained. The sections were then stained with Van Gieson's Stain and observed with a light microscope (Nikon Microphot FXA).

Statistical Analysis

Experimental data were expressed as means ± SD and the Student's t-test or one-way analysis of variance (ANOVA) with post hoc tests was applied to comparison. Differences were considered statistically significant at $p<0.05$.

RESULTS

Characterization of CPC-BG

After setting for 24 h in a 100% relative humidity box at 37°C, XRD showed that the hardened CPC contained diffraction peaks of HA. The main peaks for the HA of a hardened CPC-BG composite, were not obviously altered and peaks for Ca_2SiO_4 could be seen in the XRD patterns of the CPC-BG composite with 10% BG (Figure 1). Moreover, the presence of Ca_2SiO_4 and Ca_3SiO_5 within the CPC-BG composite containing 20% BG was also confirmed by XRD patterns.

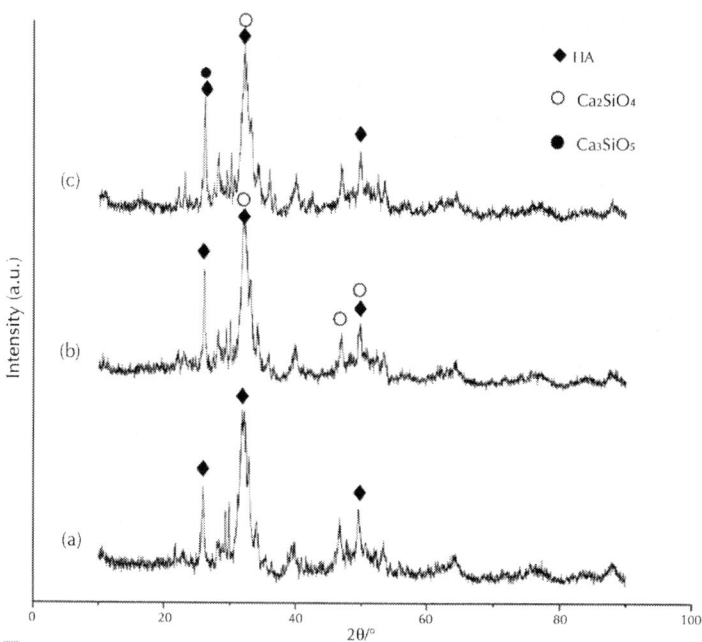

Figure 1: X-ray diffraction patterns of the CPC and CPC-BG composite specimens after setting for 24 h. (a) CPC. (b) CPC+10% BG. (c) CPC+20% BG.

The SEM micrographs for the cross section of the CPC and CPC-BG composite specimens (10% and 20% BG) showed that the CPC-BG composite specimens closely combined with each other and showed more compact microstructure than CPC alone, while the CPC formed a clay-like structure with many micropores, after setting for 24 h in a 100% relative humidity box at 37°C (Figure 2a–2c). Moreover, the microstructure of the hardened CPC-BG composite specimens became more compact with the increased BG content.

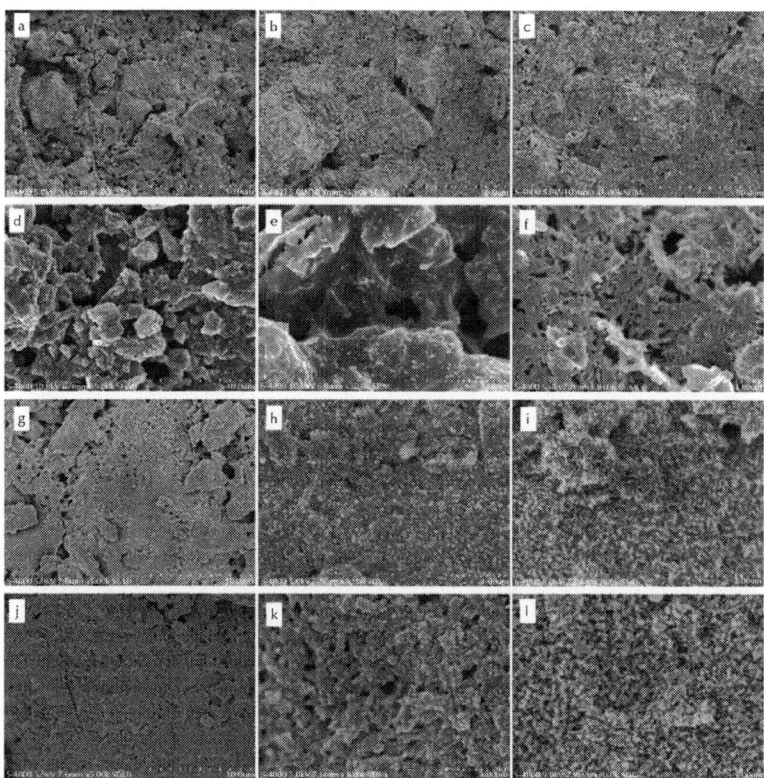

Figure 2: SEM micrographs of cross-sections of CPC and CPC-BG composite specimens that were hardened for 24 h (a–c) and specimens immersed in SBF for different times (d–l). (a) CPC. (b) CPC+10% BG. (c) CPC+20% BG. (d, e) CPC after 7 days. (f) CPC after 14 days. (g, h) CPC+10% BG after 7 days. (i) CPC+10% BG after 14 days. (j, k) CPC+20% BG after 7 days. (l) CPC+20% BG after 14 days.

Setting Time, Injectability and Compressive Strength

For all CPC-BG composite pastes, the setting times were prolonged as the content of BG increased; the time increased from 21 min to 25 min when the weight ratio of BG varied from 10% to 20% at a P/L ratio of 2.0 g/ml (Table 1).

Table 1: The setting time of the cement pastes (P/L = 2.0 g/ml)

	CPC	CPC+10% BG	CPC+20% BG
Setting time (min)	15 ± 0.8	21 ± 1.4	25 ± 0.9

Doi:10.1371/journal.pone.0062570.t001

The injectability of the CPC-BG composite paste was significantly improved compared with the injectability of the CPC paste (Figure 3a). Moreover, the CPC-BG composite paste did not give any demixing when the weight ratio of BG increased from 10% to 20% due to the filter-pressing effect during extrusion through the syringe. Furthermore, the injectability of the CPC-BG composite paste rose with an increase in BG content.

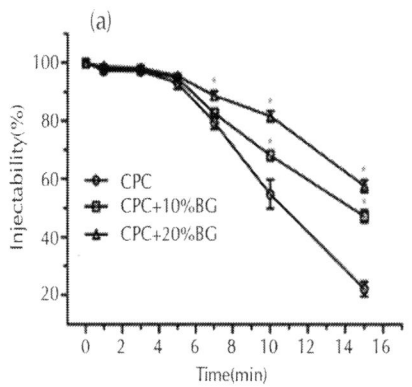
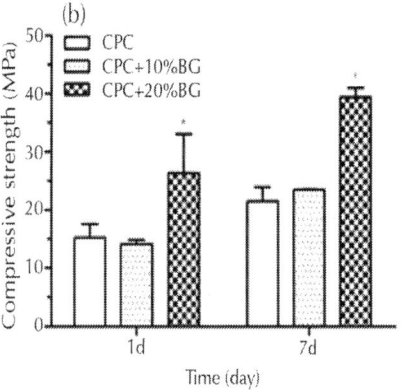

Figure 3: Injectability of the CPC and CPC-BG composite pastes versus setting time (P/L = 2.0 g/ml) (a) and compressive strength of the CPC and CPC-

BG composite specimens after setting for 1 and 7 days (P/L = 2.0 g/ml) (b). An asterisk (*) indicates that the injectability and compressive strength of the CPC-BG composite specimens were significantly different from those of CPC ($p<0.05$).

After setting in a 100% relative humidity box at 37°C for 1 and 7 days, the compressive strength of CPC and CPC-BG composite specimens increased with prolonged time during setting (Figure 3b). Furthermore, the compressive strength of the CPC-BG composites rose with increasing BG weight ratio. The composite with 20% BG exhibited the highest compressive strength after setting for 7 days. The compressive strength of CPC-BG composites (20%) reached 26 MPa at 1 day and 40 MPa at 7 days compared with only 15 MPa and 22 MPa of CPC at the same time points, respectively. There were significant differences between CPC and CPC-BG composite (20%) specimens at day 1 and day 7 ($p<0.05$).

Bioactivity and Degradation in SBF

After soaking for 7 days, the nano-sized aggregates of the bone-like apatite appeared on both surfaces of both CPC and CPC-BG composite specimens (Figure 2d–2l). With longer immersion time, the amount and grain size of apatite particles on the CPC and CPC-BG composite surfaces increased; thus, the apatite layers increased in density. However, the apatite aggregates on the CPC-BG composite surface were larger in number and denser than those on CPC surface after immersion for 7 and 14 days. Moreover, many crystals formed agglomerates and further congregated to form a layer on the surface of the CPC-BG composite and there was a noticeable increase in the density of the apatite structures with the increase of BG content.

EDS indicated that the surfaces of the CPC-BG composites (10% and 20%) consisted mainly of a calcium phosphate with Ca/P ratios of approximately 1.56 and 1.53, respectively with Si and some Na and Mg ions from soaking in the SBF solution for 7 days. Conversely, the surface of CPC consisted of a calcium phosphate with a Ca/P ratio of approximately 1.67, and it contained no Si but did have some Na and Mg ions from the SBF solution (Figure 4).

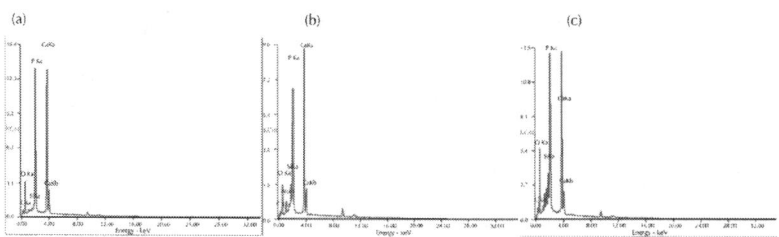

Figure 4: EDS analysis of the CPC and CPC-BG composite specimens immersed in SBF for 7 days. (a) CPC. (b) CPC+10% BG. (c) CPC+20% BG.

The degradation rates of the specimens were characterized by their weight loss ratios in SBF solution. After soaking in SBF solution at 37°C for 28 days, the degradation rates of all CPC-BG composite specimens were significantly higher than the degradation rates of CPC specimens from 7 to 14 days ($p<0.05$).The degradation rates of CPC-BG composite specimens (20%) were significantly higher from 21 to 28 days compared with the degradation rates of CPC specimens ($p<0.05$) (Figure 5). Furthermore, the degradation rate rose with an increase in BG content.

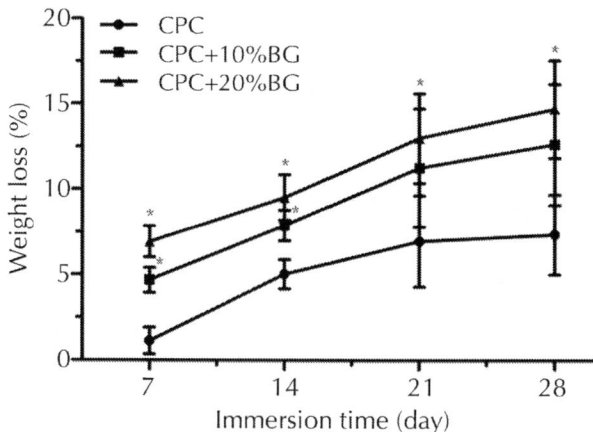

Figure 5: Weight-loss ratios of the CPC and CPC-BG composite specimens immersed in SBF for various periods. An asterisk (*) indicates that the weight-loss ratio of the CPC-BG composite specimens was significantly different than that of CPC ($p<0.05$).

Cell Attachment, Proliferation, Morphology and Differentiation

The MTT assay was adopted to assess the number of cells that adhered to the various biomaterials because OD absorbance values can be used as an indicator of the number of cells. There were no significant differences at 4 h for all the specimens (Figure 6a). However, the OD values of the CPC-BG composite specimens (20%) were significantly higher than the OD values of the CPC specimens ($p<0.05$) after a period of 8 h. There was no significant difference between CPC and CPC-BG composite (10%) although the OD values of CPC-BG composite (10%) were higher than the OD values of cells on CPC. In addition, the OD values of the CPC-BG composite rose with an increase in BG content and the highest OD values were obtained in the composite with 20% BG after 8 h.

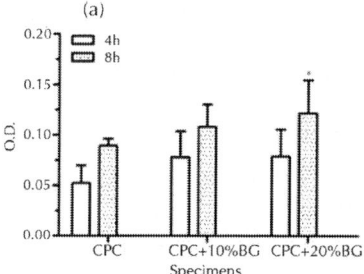 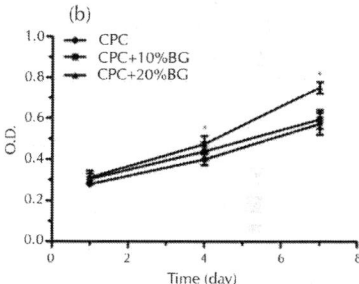

Figure 6: Cell attachment on CPC and CPC-BG composite specimens at 4 h and 8 h (a), Cell viability on CPC and CPC-BG composite specimens after 1, 4 and 7 days (b). An asterisk (*) indicates that the cell attachment and proliferation of CPC-BG composite specimens were significantly different from those of CPC ($p<0.05$).

The viability of osteoblasts cultured on CPC and CPC-BG composite specimens was assessed with the MTT assay because the OD values can provide an indication of cell growth and proliferation on various biomaterials. The OD values of the CPC-BG composite specimens (20%) were significantly higher than the OD values of the CPC specimens after 4 and 7 days ($p<0.05$), indicating that the CPC-BG composite specimens promoted cell growth and facilitated proliferation with no cytotoxic effect on cells compared with CPC specimens (Figure 6b).

Cells firmly attached and spread well on the surface of CPC and CPC-BG composite specimens with morphologically normal appearance after for 1 day of culture (Figure 7a, b). As number of cells increased, the cells extended, spread well and exhibited intimate attachment to the surfaces of the CPC and CPC-BG composite specimens with cytoplasmic extensions after 4 days (Figure 7c, d). Cell-to-cell junctions appeared in the SEM images. The number of cells on the CPC-BG composite specimens was greater than the amount on the CPC specimens.

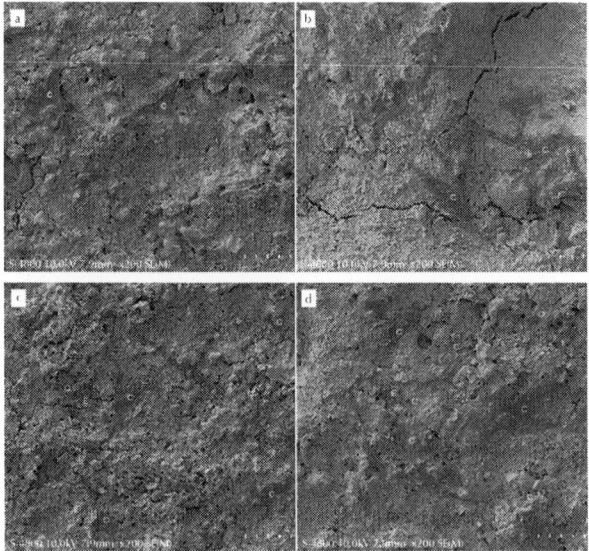

Figure 7: SEM micrographs of the morphological features of cells cultured on the CPC and CPC-BG composite specimens for 1 and 4 days. Images of cells cultured on (a) CPC and (b) CPC-BG composite specimens for 1 day. Images of cells cultured on (c) CPC and (d) CPC-BG composite specimens for 4 days. C: cells. E: Cytoplasmic extensions of the cells. J: cell-cell junctions.

Cell differentiation was evaluated by testing the ALP activity of rat osteoblasts cultured on CPC and CPC-BG composite specimens after 4 and 7 days. The ALP activity of cells cultured on the CPC-BG composite (20%) was significantly higher than the ALP activity on the CPC and TCPS controls ($p<0.05$) after 4 and 7 days (Figure 8). Moreover, there was no significant difference among CPC, CPC-BG composite (10%), and TCPS control.

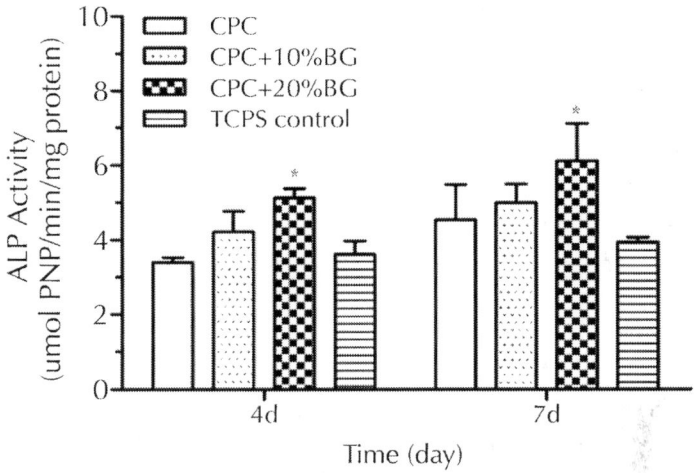

Figure 8: Alkaline phosphatase activity (ALP) of cells cultured on the CPC and CPC-BG composite specimens for 4 and 7 days with tissue-cultured polystyrene (TCPS) as the control. An asterisk (*) indicates that the ALP activities of cells cultured on CPC-BG composite specimens were significantly different from those of CPC and the TCPS control ($p<0.05$).

Macroscopic Evaluation

After 4 weeks' implantation, macroscopic observations of CPC and CPC-BG composites implanted into the bone defects of rabbit lateral femoral condyles showed that the implants exhibited no foreign body reaction, no inflammation and no necrosis in vivo and were incorporated well with surrounding tissue. All CPC-BG composite specimens were covered with a tissue layer that was indistinguishable from surrounding tissue (Figure 9a, b). Conversely, CPC specimens were covered with a thinner tissue layer that could be distinguished by macroscopic evaluation (Figure 9c). With the increase of the implantation period to 12 weeks, the volume of CPC-BG composite specimens decreased accompanied by a simultaneous bone ingrowth from the periphery inwards due to the degradation of the CPC-BG composite. The boundaries between normal surrounding tissue and the specimens were indistinct, and the newly formed bone could not be distinguished from normal bone (Figure 9d, e). Conversely, CPC specimens did not show observable variations in size from the original implantation after 12 weeks of degradation and the bone ingrowth mainly occurred at the native bone margins and the defect periphery (Figure 9f).

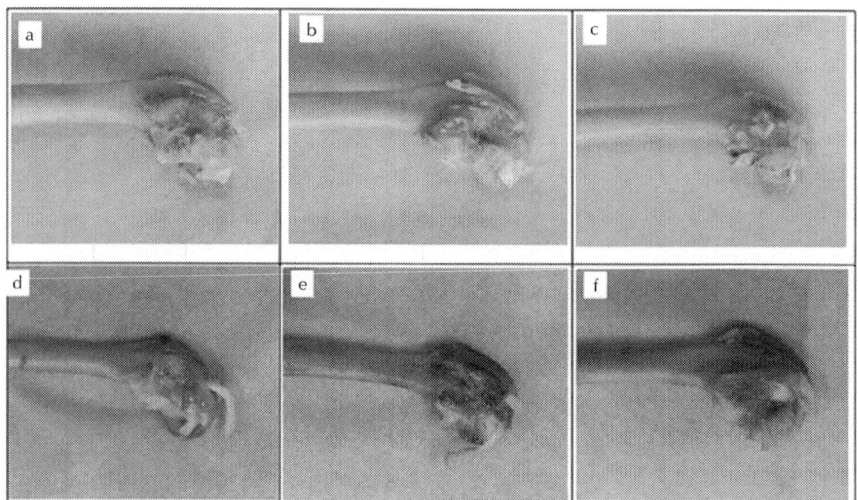

Figure 9: Macroscopic evaluation of CPC-BG composite and CPC specimens implanted into bone defects of rabbits for 4 and 12 weeks. (a) CPC+10% BG for 4 weeks. (b) CPC+20% BG for 4 weeks. (c) CPC for 4 weeks. (d) CPC+10% BG for 12 weeks. (e) CPC+20% BG for 12 weeks. (f) CPC specimens for 12 weeks.

Micro-CT Analysis

3D reconstruction images of residual material of the CPC and CPC-BG composite specimens after implantation for 4 and 12 weeks were adopted to assess the in vivo resorption of the implants (Figure 10a). After a prolonged implantation time from 4 to 12 weeks, the surface morphologies of the CPC-BG composite specimens exhibited many differences from their original appearances. After 12 weeks' implantation, a porous surface structure was obtained. With increased implantation time, the pore size formed by degradation became larger and the volumes of the CPC-BG composite specimens decreased. Conversely, the CPC specimens showed rare variations in appearance and volume and pore formation was mainly found at the outermost edge of the implants at 12 weeks after implantation. Figure 10b displays the bone ingrowth into the implants at 4 to 12 weeks. At 4 weeks, there was a small amount of newly formed bone at the interface of the CPC-BG composite specimens, and more extensive bone ingrowth occurred throughout the cross-section of specimens at 12 weeks after implantation. However, bone formation in the CPC group was mostly observed at the defect periphery at 4 weeks and only a small amount of new bone tissue was found within the implants 12 weeks after implantation (Figure 10b). From 4 to 12 weeks, the RMVF of CPC-BG (20%) composite specimens decreased from 81.16±2.66% to 66.18±1.32% whereas the RMVF of CPC remained at 84.42±3.20% after 12 weeks of implantation, revealing that the in vivo degradation of the CPC-BG composites was much higher than that of CPC ($p<0.05$) (Figure 11a). The BVF was applied to evaluate the newly formed bone more precisely (Figure 11b). The BVF of each CPC-BG composite specimen was significantly higher than the BVF of each CPC specimen at both 4 and 12 weeks ($p<0.05$).

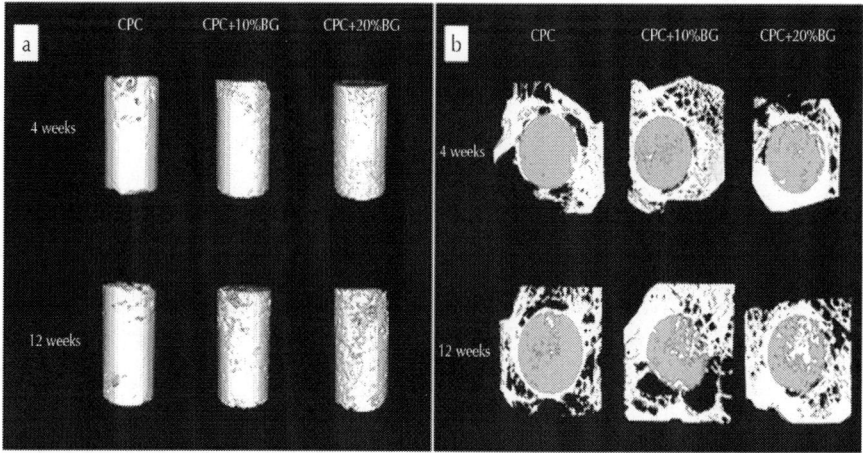

Figure 10: Three-dimensional reconstruction using micro-CT analysis. (a) residual material of the CPC-BG composite and CPC. (b) cross-sectional images of rabbit femur after implantation for 4 and 12 weeks.

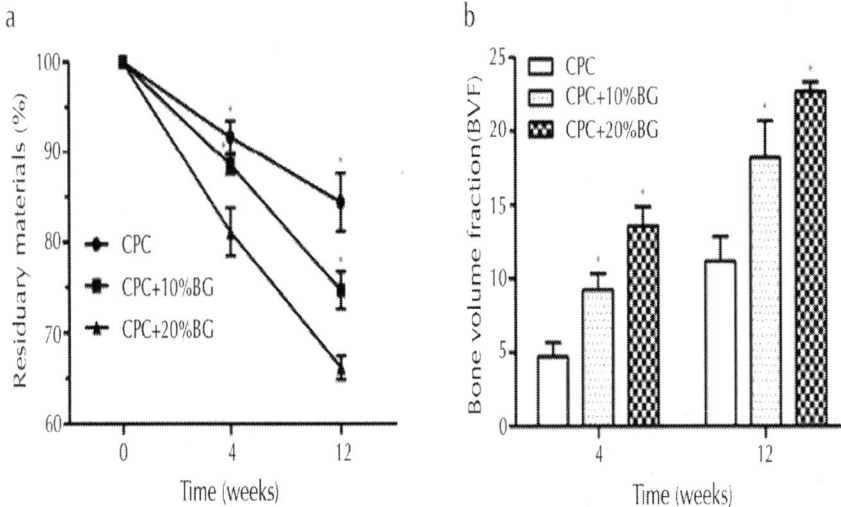

Figure 11: Quantitative analysis of residual material and new bone formation from micro-CT images. (a) residual material. (b) new bone formation for 4 and 12 weeks.

These results confirmed that CPC-BG showed excellent biocompatibility, degradability and osteogenesis, and CPC-BG exhibited greater bone-forming efficiency than CPC.

Histological Analysis

After 4 weeks' implantation, the CPC-BG composite implant was encapsulated by the surrounding bone tissue and the boundary between the implant and host bone was detectable. The material had started to degrade from the edge of the implant, and new bone tissues had grown into the pores formed by the degradation of the CPC-BG composite implant. The new bone was in direct contact with the surface of the implant (Figure 12a–d). After 12 weeks' implantation, bone ingrowth had occurred in many areas of the implant and the amount of newly formed bone in those defects had increased dramatically together with the resorption of the CPC-BG composite implant. Direct contact between the new bone and the CPC-BG composite implant increased from 4 weeks to 12 weeks (Figure 12g–j). For CPC, resorption of CPC rarely occurred, and new bone tissue formed only rarely at the interface of the implant after 4 weeks (Figure 12e, f). After 12 weeks' implantation, there was only marginal degradation in the CPC implant and no fibrous layer appeared between the bone and the implant surface (Figure 12k, l), which is in accordance with the results of the macroscopic evaluation and the micro-CT analysis. These results confirmed that the CPC-BG composite showed excellent biocompatibility, biodegradability and osteogenesis, and that the CPC-BG exhibited significant advantages over CPC.

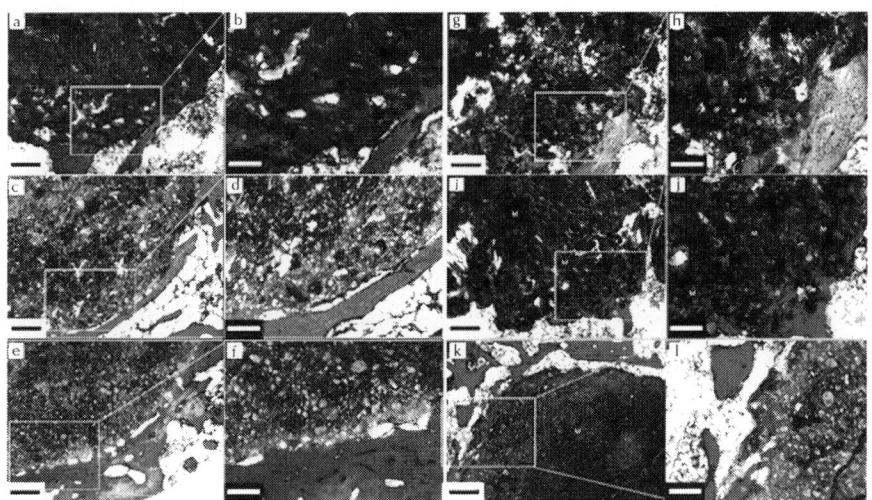

Figure 12: Van Gieson's-stained sections of CPC-BG composite and CPC specimens that were harvested at 4 (a–f) and 12 (g–l) weeks after implantation. (a, b) CPC+20% BG for 4 weeks. (c, d) CPC+10% BG for 4 weeks. (e, f) CPC for 4 weeks. (g, h) CPC+20% BG for 12 weeks. (i, j) CPC+10% BG for 12 weeks. (k, l) CPC for 12 weeks. M: materials. B: bone formation. Scale bar: 400 μm (black), 160 μm (white).

DISCUSSION

A major drawback of orthopedic implant materials such as hydroxyapatie (HA) ceramic in current use is their hardened form, which requires the surgeon to drill the surgical site around the graft or to carve the implant into the desired shape and often leads to increased bone loss, trauma and surgical time [21]. With the emergence of minimally invasive surgery techniques, materials with self-setting properties have drawn more attention [22], [23]. In the present study, a novel injectable CPC-BG composite has been developed that can be mixed with the cement liquid (potassium phosphate buffers) to form a paste that can be applied to fill irregular bone cavities during surgery via injection especially where shaping and contouring for esthetics are needed. The material exhibited a retarded setting time and, consequently, improved injectability due to the addition of BG as compared with CPC paste. However, in clinical applications, the cement must be applied within the initial setting time and be extruded during the operation. Compared with CPC paste, which sets in <10 min, CPC-BG composite paste's extra setting time (21 min for 10 wt% BG or 25 min for 20 wt% BG) gives an advantage to surgeons or dentists by allowing more time for the operation before the paste set.

As the cement is used for bone repair, the mechanical properties of the hardened cement are another important index [24]. Many researchers have worked to improve the mechanical strength of CPC [7], [18], [25]. For instance, the mechanical strength of DCPD/C_3S cement is improved by mixing $CaHPO_4 \cdot 2H_2O$ (DCPD) and Ca_3SiO_5 (C_3S) mainly due to the filler effect of the calcium silicate's hydrated phase and the chemical interaction between the CPC matrix and C_3S particles. However, the setting time of the DCPD/C_3S cement with 40 wt% C3S was prolonged to approximatley 60 min with increased C_3S amount, which would enhance the chance of the cement being washed out by physiological liquid during clinical application [26]. In the present study, the compressive strength of CPC-BG composite specimens was significantly higher than the compressive strength of CPC, and the CPC-BG composite specimens with 20 wt% BG possessed the highest compressive strength, at approximaltely 26 MPa and 40 MPa after setting for 1 and 7 days, respectively. This high strength may be due to the lower setting rate and consequently more compact microstructure of the CPC-BG composite (Figure 2). Moreover, Si ions tend to inhibit grain the growth of HA crystals [27], which further contributes to the tendency of the CPC-BG composite to form a compact and homogeneous microstructure. However, the exact interaction mechanism between CPC and BG need to be further illuminated.

For bioactive substitution materials, it is important to induce a bone-like apatite layer on the surface that can form chemical bonds to bone tissue at the early stage of the implantation [28]. Bioglass has been reported to form bone-like apatite on its surface after being soaked in SBF [29]. In the present study, the number and grain size of apatite aggregates on CPC and CPC-BG composite surfaces increased with prolonged immersion time. However, the number of apatite aggregates on the surface of the CPC-BG composite was larger than the amount on the CPC surface after immersion for 7 and 14 days. Moreover, this apatite layer grew more densely when there was a greater content of BG in the CPC-BG composite, indicating that the bioactivity of the CPC-BG composite was improved. These results suggest that the addition of BG could result in a bioactive composite with controllable bioactivity. It is well known that the formation of apatite on the surface of CPC mainly relies on supersaturation of Ca^{2+} and PO_4^{3-} ions and that the process of deposition proceeding at a low rate leads to absent homogeneous apatite layer formation at the early stages of the implantation [30]. With the addition of BG in the composite, a SiO^{2-} rich gel layer forms on the surface of the BG and plays an important role in the formation of a CaO-P_2O_5-rich film, which provides favorable sites for nucleation of apatite crystals that form the apatite layer in the simulated body environment. As the content of BG increases, the CPC-BG composite provides more nucleation sites for the apatite crystals, resulting in the formation of a homogeneous apatite layer on the surface of the CPC-BG composite. The assumption that the CPC-BG composite forms a stronger bond with the surrounding bone tissue than CPC needs to be further confirmed by in vivo studies.

The biomaterial should be degradable and gradually replaced by newly formed bone tissue [31]. Proper degradation in a physiological environment is one of the most important characteristics of in bone repair applications. The degradation rate of CPC, which is composed of TECP and DCPA, is slow [25]. In the present study, the degradation rate of CPC-BG composite was significantly faster than the degradation of CPC, which can be attributed to the higher solubility of BG. In addition, the degradation rate of the CPC-BG composite can be adjusted by controlling the BG content.

The cellular responses to biomaterials can be influenced by surface characteristics of the biomaterials in vitro. Ideally, bioactive materials should interact actively with cells and stimulate cell growth [32]. A first effort to culture rat osteoblasts on the novel CPC-BG composite was performed to evaluate cell attachment, proliferation and differentiation. Because the cell attachment stage is the initial stage of interaction between the cells and the biomaterial, its quality will directly affect cell growth, morphology, proliferation and differentiation [34]. Cells adhered better to the surface of the CPC-BG composite specimen than CPC after 8 h, indicating that the CPC-BG composite was superior over CPC in promoting cellular attachment. Furthermore, cells adhered best to the CPC-BG composite with 20 wt% BG, suggesting that BG played an important role in promoting cellular attachment. It has been known that surface silanols (Si-OH) of BG, which generate a high negative-surface-charge density, contribute to the strong irreversible adsorption of serum proteins [35]. The enhancement of cell attachment was mostly likely associated with the preferential adsorption of serum proteins such as fibronectin onto BG, an intermediary step preceding cell attachment to the biomaterial surface [36].

Both CPC and BG have been shown to be biocompatible in previous in vitro and in vivo studies [33]. The novel CPC-BG composite exhibited a positive cellular behavior and was shown to be cyto-compatible with no obvious negative effects on cellular viability.

ALP activity was routinely used as an early marker of osteoblast differentiation in vitro. Fetal osteoblastes were shown to differentiate in a 45S5 Bioglass® conditioned medium in the absence of osteogenic supplements [12]. 45S5 BG exhibited a significant effect on the early differentiation of marrow stromal cells into osteoblast-like cells [10]. Interestingly, it has been reported that this effect on differentiation of rat MSCs, was not observed in human MSCs when cells were grown on BG or with BG dissolution products [37]. Similarly, additional findings showed that BG-supplemented materials did not affect ALP of human MSCs in vitro but they did elicit a marked increase in bone formation in vivo where a complex mixture of cells and growth factors was present. This difference indicates that BG supports bone formation through a more complex mechanism than direct stimulation of MSC differentiation [38]–[40]. Recently, it was reported that PLGA-S-BG composites supported BMP-mediated osteogenesis of human BMSCs, which reflects the good osteoinductive properties of BG [41]. In the present study, our results showed that the CPC-BG composite promoted cellular differentiation and possessed excellent bioactivity when BG was added as a constituent. Thus, to better understand the effect of BG on bone cells, a more thorough study on the human MSC response to BG or BG-supplemented materials in vitro is necessary.

Biocompatibility is a factor relevant to the response of cells that are in contact with the biomaterial, and it has been reported that the surface of biomaterials may affect the behavior and morphology of cells cultured on their surface [42]. SEM results for cell morphology confirmed that cells attached and spread on the surfaces of both the CPC-BG composite and CPC. The observed cell-to-cell junctions revealed that both the CPC-BG composite and CPC were suitable for cell attachment and growth. At day 4, the number of cells on the CPC-BG composite appeared to be more than the number on the CPC specimens, which can be attributed to strong irreversible adsorption of serum proteins onto BG.

Moreover, cellular responses to biomaterials, such as cell attachment, proliferation and differentiation, depend not only on the surface morphology but also on the chemical composition of the biomaterial [43], which plays a crucial part in determining the cell-material interaction for biomaterials by influencing the quantity of ions released from the biomaterial [44]. Previous studies have demonstrated that ion dissolution products containing Ca and Si that were released from BG can stimulate cell attachment, proliferation, differentiation and mineralization [10]–[14]. In the present study, dissolution of the CPC-BG composite provided a Ca- and Si-rich environment to stimulate cell growth, proliferation and differentiation.

In the in vivo study, the macroscopic evaluation results showed that both the CPC-BG composite and CPC implants exhibited no obvious inflammatory response, rejection or necrosis in the adjacent host tissue and they incorporated well with the surrounding tissue. With the prolonged time to 12 weeks, the boundaries between CPC-BG composite specimens and normal surrounding tissue were indistinct due to the degradation of specimens and subsequent ingrowth of new bone. Conversely, the CPC specimens exhibited few variations in size after 12 weeks implantation. It is well known that resorption of the bone-substitute material is required in the replacement of bone tissue because bone ingrowth into the defect area requires the liberation of the space [45]. In this study, the precise evaluation of in vivo degradation and newly formed bone was confirmed with micro-CT analysis. The in vivo resorption increased with prolonged implantation time. There was remarkably higher in vivo degradation of CPC-BG composites, which substantially influenced bone formation. As the implantation time increased, new bone was regenerated and gradually penetrated into the implant accompanied by the resorption of the CPC-BG composite implant. It is believed that chemical dissolution was the main way of resorption for the implant during the early period of implantation because it changed and enlarged the microstructure of the implant, which could facilitate cell-mediated resorption later on. The increased degradation of the CPC-BG composite implant might be related to the increased dissolution of BG after contact with fluids [40]. Additionally, ionic dissolution products of BG have been reported to beneficially affect osteogenesis by formation of a hydroxycarbonate apatite (HCA) layer and promotion of bone growth [11], [12]. Moreover, it has been suggested that BG has a stimulatory effect on neovascularization [39], [40], which, together with the osteopromotive properties of BG, might further influence bone formation when using a CPC-BG composite as the bone-substitute material. BVF gradually increased while the volume of the CPC-BG composite specimens continued to decrease over time, which indicates that cell-mediated resorption occurred. Direct and intimate contact appeared at the interfaces of both CPC-BG composite and CPC specimens. However, a quantitative analysis showed that the BVF values for the CPC-BG composite specimens were greatly higher than the values for CPC. Moreover, more extensive bone in growth occurred throughout the cross-sections of the CPC-BG composite specimens, which indicates that more effective osteogenesis and oseointegration had occurred at the defect area; these indicators are considered to be critical to firmly anchor the implant in place [46]. Histological evaluation revealed that the CPC-BG composite specimens were encapsulated by the surrounding bone tissue, and the new bone was in direct contact with the implant 4 week after implantation. New bone tissues formation increased considerably both in and along the implant together with the resorption of the CPC-BG composite implant 12 weeks after implantation. The implant of CPC-BG composite implant formed tight and direct bonding with the surrounding host bone without the intervention of soft tissue, in accordance with macroscopic evaluation and micro-CT results. These results confirmed that the CPC-BG composites exhibited not only faster biodegradability but also enhanced and more effective osteogenesis and osteointegration at the defect area; these are significant advantages over CPC. However, our study was carried out under optimal conditions. Considering the differences in bone metabolism for healthy bone compared with those under compromised conditions, the biological performance of CPC-BG composite should be tested for compromised conditions such as osteoporosis. Moreover, long-term studies of CPC-BG composites for bone regeneration should be carried out. As an ideal implanted biomaterial candidate for bone regeneration, the CPC-BG composites discussed herein presented good biocompatibility and osteoconductive properties and induced bone ingrowth into the implant. In addition, the material was shown to be resorbable and it is replaced by new bone in a creeping substitution manner. Finally, the material can be handled as a paste and set in situ within a comfortable time. Our findings suggested that CPC-BG is a potential bioactive composite material for bone regeneration in future clinical situations.

CONCLUSIONS

A novel injectable CPC-BG composite was first fabricated and characterized by incorporating BG into CPC and showed a prolonged setting time with improved injectability. The mechanical strength of the CPC-BG composite was significantly enhanced over CPC. Furthermore, incorporation of BG into the CPC appeared to significantly improve the bioactivity and degradability. The CPC-BG composite promotes the attachment, proliferation and differentiation of rat osteoblasts and exhibits excellent biocompatibility with no negative effects on cell morphology or viability. Macroscopic observations of CPC-BG composite implants exhibited no obvious inflammatory response, rejection or necrosis, and the implants incorporated well with the surrounding tissue in vivo. Micro-CT and histological evaluations confirmed that the CPC-BG composite implants exhibited more effective osteogenesis and osteointegration at the defect area than CPC with good biocompatibility and biodegradability. In conclusion, this novel injectable biomaterial with improved properties by incorporation of BG into CPC exhibits promising prospects for bone regeneration.

AUTHOR CONTRIBUTIONS

Conceived and designed the experiments: LY ZXW WL. Performed the experiments: YL KZ YFT ZC JC YZ JWW SL LK. Analyzed the data: LY YL. Contributed reagents/materials/analysis tools: KZ YFT ZC. Wrote the paper: LY ZXW.

REFERENCES

1.   Wu F, Wei L, Guo H, Liu CS (2008) Self-setting bioactive calcium-magnesium phosphate cement with high strength and degradability for bone regeneration. Acta Biomaterialia 4: 1873–1884. doi: 10.1016/j.actbio.2008.06.020

2.   Komaki H, Tanaka T, Chazono M, Kikuchi T (2006) Repair of segmental bone defects in rabbit tibiae using a complex of beta-tricalcium phosphate, type I collagen, and fibroblast growth factor-2. Biomaterials 27: 5118–5126. doi: 10.1016/j.biomaterials.2006.05.031

3.   Link DP, van den Dolder J, van den Beucken JJ, Wolke JG, Mikos AG, et al. (2008) Bone response and mechanical strength of rabbit femoral defects filled with injectable CaP cements containing TGF-b1 loaded gelatin microspheres. Biomaterials 29: 675–682. doi: 10.1016/j.biomaterials.2007.10.029

4.   Brown WE, Chow LC (1986) A new calcium phosphate water setting cement. In: Brown PW, editor. Cements research progress. Westerville, OH: American Ceramic Society p. 352–379.

5.   Julien M, Khairoun I, LeGeros RZ, Delplace S, Pilet P, et al. (2007) Physico-chemical–mechanical and in vitro biological properties of calcium phosphate cements with doped amorphous calcium phosphates. Biomaterials 28: 956–965. doi: 10.1016/j.biomaterials.2006.10.018

6.   Friedman CD, Costantino PD, Takagi S, Chow LC (1998) Bonesource hydroxyapatite cement: a novel biomaterial for craniofacial skeletal tissue engineering and reconstruction. J Biomed Mater Res 43B: 428–432. doi: 10.1002/(sici)1097-4636(199824)43:4<428::aid-jbm10>3.0.co;2-0

7. Moreau JL, Xu HHK (2009) Mesenchymal stem cell proliferation and differentiation on an injectable calcium phosphate- Chitosan composite scaffold, Biomaterials. 30: 2675–2682. doi: 10.1016/j.biomaterials.2009.01.022

8. Xu HHK, Zhao L, Detamore MS, Takagi S, Chow LC (2010) Umbilical cord stem cell seeding on fast-resorbable calcium phosphate bone cement. Tissue engineering: Part A. 16: 2743–2753. doi: 10.1089/ten.tea.2009.0757

9. Bodde E, Boerman O, Russel F, Mikos A, Spauwen P, et al. (2008) The kinetic and biological activity of different loaded rhBMP-2 calcium phosphate cement implants in rats. J Biomed Mater Res A 87A: 780–791. doi: 10.1002/jbm.a.31830

10. Bosetti M, Cannas M (2005) The effect of bioactive glasses on bone marrow stromal cells differentiation. Biomaterials 26: 3873–3879. doi: 10.1016/j.biomaterials.2004.09.059

11. Heikkila JT, Mattila KT, Andersson OH, Knuuti J, Yli-Urpo A, et al. (1995) Behavior of bioactive glass in human bone. Bioceramics 8: 35–41.

12. Tsigkou O, Jones JR, Polak JM, Stevens MM (2009) Differentiation of fetal osteoblasts and formation of mineralized bone nodules by 45S5 Bioglass® conditioned medium in the absence of osteogenic supplements. Biomaterials 30: 3542–3550. doi: 10.1016/j.biomaterials.2009.03.019

13. Jell G, Notingher I, Tsigkou O, Notingher P, et al. (2008) Bioactive glass-induced osteoblast differentiation: A noninvasive spectroscopic study. J Biomed Mater Res A 86A: 31–40. doi: 10.1002/jbm.a.31542

14. Christodoulou I, Buttery LD, Saravanapavan P, Tai G, Hench LL, et al. (2005) Dose- and time-dependent effect of bioactive gel-glass ionic-dissolution products on human fetal osteoblast-specific gene expression. J Biomed Mater Res B Appl Biomater 74B: 529–537. doi: 10.1002/jbm.b.30249

15. Radin S, Reilly G, Bhargave G, Leboy PS, Ducheyne P (2005) Osteogenic effects of bioactive glass on bone marrow stromal cells. J Biomed Mater Res A 73: 21–29. doi: 10.1002/jbm.a.30241

16. Renno ACM, van de Watering FCJ, Nejadnik MR, Crovace MC, et al. (2013) Incorporation of bioactive glass in calcium phosphate

cement: An evaluation. Acta Biomaterialia 9: 5728–5739. doi: 10.1016/j.actbio.2012.11.009

17. Guo H, Su JC, Wei J, Kong H, Liu CS (2009) Biocompatibility and osteogenicity of degradable Ca-deficient hydroxyapatite scaffolds from calcium phosphate cement for bone tissue engineering. Acta Biomaterialia 5: 268–278. doi: 10.1016/j.actbio.2008.07.018

18. Huan ZG, Chang J (2009) Calcium-phosphate-silicate composite bone cement: self-setting properties and in vitro bioactivity. J Mater Sci: Mater Med 20: 833–841. doi: 10.1007/s10856-008-3641-9

19. Kokubo T, Takadama H (2006) How useful is SBF in predicting in vivo bone bioactivity. Biomaterials 27: 2907–2915. doi: 10.1016/j.biomaterials.2006.01.017

20. Wada Y, Kataoka H, Yokose S, Ishizuya T, et al. (1998) Changes in osteoblast phenotype during differentiation of enzymatically isolated rat calvaria cells. Bone 22: 479–485. doi: 10.1016/s8756-3282(98)00039-8

21. Laurencin CT, Ambrosio AMA, Borden MD, Cooper JA (1999) Tissue engineering: orthopedic applications. Annu Rev Biomed Eng 1: 19–46. doi: 10.1146/annurev.bioeng.1.1.19

22. Combes C, Tadier S, Galliard H, Girod-Fullana S, et al. (2010) Rheological properties of calcium carbonate self-setting injectable paste. Acta Biomaterialia 6: 920–927. doi: 10.1016/j.actbio.2009.08.032

23. Chen ZG, Liu HY, Liu X, Cui FZ (2011) Injectable calcium sulfate/mineralized collagen-based bone repair materials with regulable self-setting properties. J Biomed Mater Res A 99: 554–563. doi: 10.1002/jbm.a.33212

24. Zhang JT, Tancret F, Bouler JM (2011) Fabrication and mechanical properties of calcium phosphate cements (CPC) for bone substitution. Mater Sci Eng C 31: 740–747. doi: 10.1016/j.msec.2010.10.014

25. Lu JX, Wei J, Yan YG, Li H, et al. (2011) Preparation and preliminary cytocompatibility of magnesium doped apatite cement with degradability for bone regeneration. J Mater Sci: Mater Med 22: 607–615. doi: 10.1007/s10856-011-4228-4

26. Lin Q, Lan XH, Li YB (2010) Anti-washout carboxymethyl chitosan modified tricalcium silicate bone cement: preparation, mechanical properties and in vitro bioactivity. J Mater Sci: Mater Med 21: 3065–3076. doi: 10.1007/s10856-010-4160-z

27. Huan ZG, Chang J (2009) Novel bioactive composite bone cements based on the β-tricalcium phosphate-monocalcium phosphate monohydrate composite cement system. Acta Biomaterialia 5: 1253–1264. doi: 10.1016/j.actbio.2008.10.006

28. Kobayashi M, Nakamura T, Shinzato S, Mousa WF, Nishio K, et al. (1999) Effect of bioactive filler content on mechanical properties and osteoconductivity of bioactive bone cement. J Biomed Mater Res 46: 447–457. doi: 10.1002/(sici)1097-4636(19990915)46:4<447::aid-jbm2>3.0.co;2-p

29. Miguel BS, Kriauciunas R, Tosatti S, Ehrbar E, et al. (2010) Enhanced osteoblastic activity and bone regeneration using surface-modified porous bioactive glass scaffolds. J Biomed Mater Res A 94A: 1023–1033. doi: 10.1002/jbm.a.32773

30. He Q, Chen H, Huang L, Dong J, Guo D, et al. (2012) Porous surface modified bioactive bone cement for enhanced bone bonding. PLoS One 7: e42525. doi: 10.1371/journal.pone.0042525

31. Hu GF, Xiao LW, Fu H, Bi DW, et al. (2010) Study on injectable and degradable cement of calcium sulphate and calcium phosphate for bone repair. J Mater Sci: Mater Med 21: 627–634. doi: 10.1007/s10856-009-3885-z

32. Chen QZ, Efthymiou A, Salih V, Boccaccini AR (2008) Biogalss® -derived glass-ceramic scaffolds: Study of cell proliferation and scaffold degradation in vitro. J Biomed Mater Res A 84A: 1049–1060. doi: 10.1002/jbm.a.31512

33. Saboor A, Rabiee M, Mutarzadeh F, Sheikhi M, Tahriri M, et al. (2009) Synthesis, characterization and in vitro bioactivity of sol-gel-derived SiO_2-CaO-P_2O_5-MgO bioglass. Mater. Sci 29: 335–340. doi: 10.1016/j.msec.2008.07.004

34. Low SP, Williams KA, Canham LT, Voelcker NH (2006) Evaluation of mammalian cell adhesion on surface modified porous silicon. Biomaterials 27: 4538–4546. doi: 10.1016/j.biomaterials.2006.04.015

35. Manitha B, Nair HK, Varma AJ (2009) Triphasic ceramic coated hydroxyapatite as a niche for goat cell-derived osteoblasts for bone regeneration and repair. J Mater Sci: Mater Med 20: S251–258. doi: 10.1007/s10856-008-3598-8

36. Lee MH, Ducheyne P, Lynch L, et al. (2006) Effect of biomaterial surface properties on fibronectin-alpha(5)beta(1) integrin interaction and cellular attachment. Biomaterials 27: 1907–1916. doi: 10.1016/j.biomaterials.2005.11.003

37. Reilly GC, Radin S, Chen AT, Ducheyne P (2007) Differential alkaline phosphatase responses of rat and human bone marrow derived mesenchymal stem cells to 45S5 bioactive glass. Biomaterials 28: 4091–4097. doi: 10.1016/j.biomaterials.2007.05.038

38. Yao J, Radin S, Reilly G, Leboy P, Ducheyne P (2005) Solution-mediated effect of bioactive glass in poly(lactic-co-glycolic acid)-bioactive glass composites on osteogenesis of marrow stromal cells. J Biomed Mater Res 75A: 794–801. doi: 10.1002/jbm.a.30494

39. Leach JK, Kaigler D, Wang Z, Krebsbach PH, Mooney DJ (2006) Coating of vegf-releasing scaffold with bioactive glass for angiogenesis and bone regeneration. Biomaterials 27: 3249–3255. doi: 10.1016/j.biomaterials.2006.01.033

40. Hoppe A, Guldal NS, Boccaccini AR (2011) A review of the biological response to ionic dissolution products from bioactive glasses and glass-ceramics. Biomaterials 32: 2757–2774. doi: 10.1016/j.biomaterials.2011.01.004

41. Pamula E, Kokoszka J, Cholewa-kowalska K, Laczka M, Kantor L, et al. (2011) Degradation, bioactivity, and osteogenic potential of composites made of PLGA and two different sol-gel bioactive glasses. Ann Biomed Eng. 39: 2114–2129. doi: 10.1007/s10439-011-0307-4

42. Lee JY, Kang BS, Hicks B, Chancellor TF, Chu BH, et al. (2008) The control of cell adhesion and viability by zinc oxide nanorods. Biomaterials 29: 3743–3749. doi: 10.1016/j.biomaterials.2008.05.029

43. Chou SY, Cheng CM, LeDuc PR (2009) Composite polymer systems with control of local substrate elasticity and their effect on cytoskeletal and morphological characteristics of

adherent cells. Biomaterials 30: 3136–3142. doi: 10.1016/j.biomaterials.2009.02.037

44. Wu C, Ramaswmy Y, Zhu YF, Zheng R, Appleyard R, et al. (2009) The effect of mesoporous bioactive glass on the physiochemical, biological and drug-release properties of poly(DL-lactide-co-glycolide) films. Biomaterials 30: 2199–2208. doi: 10.1016/j.biomaterials.2009.01.029

45. van de Watering FCJ, van den Beucken JJJP, Walboomers XF, Jansen JA (2012) Calcium phosphate/poly (D, L-lactic-co-glycolic acid) composite bone substitute materials: evaluation of temporal degradation and bone ingrowth in a rat critical-sized cranial defect. Clin Oral Implants Res 23: 151–159. doi: 10.1111/j.1600-0501.2011.02218.x

46. Hennessy KM, Clem WC, Phipps MC, Sawyer AA, Shaikh FM, et al. (2008) The effect of RGD peptides on osseointegration of hydroxyapatite. Biomaterials 29: 3075–3083. doi: 10.1016/j.biomaterials.2008.04.014

Citations

CHAPTER 1

Garces, H. , Roller, J. , King'ondu, C. , Dharmarathna, S. , Ristau, R. , Jain, R. , Maric, R. and Suib, S. (2014) Formation of Platinum (Pt) Nanocluster Coatings on K-OMS-2 Manganese Oxide Membranes by Reactive Spray Deposition Technique (RSDT) for Extended Stability during CO Oxidation. Advances in Chemical Engineering and Science, 4, 23-35. doi: 10.4236/aces.2014.41004.

CHAPTER 2

Margarita J. Ramírez-Moreno, Issis C. Romero-Ibarra, José Ortiz-Landeros and Heriberto Pfeiffer (2014). Alkaline and Alkaline-Earth Ceramic Oxides for CO_2 Capture, Separation and Subsequent Catalytic Chemical Conversion, CO_2 Sequestration and Valorization, Mr. Victor Esteves (Ed.), ISBN: 978-953-51-1225-9, InTech, DOI: 10.5772/57444.

CHAPTER 3

Jongmuk Won, Dongseop Lee, Kyunguk Na, In-Mo Lee, Hangseok Choi, Physical properties of G-class cement for geothermal well cementing in South Korea, Renewable Energy, Volume 80, August 2015, Pages 123-131, ISSN 0960-1481, http://dx.doi.org/10.1016/j.renene.2015.01.067.

CHAPTER 4

G Espinosa-Paredes, a García, E Santoyo, E Contreras, and J.M Morales, Thermal Property Measurement of Mexican Geothermal Cementing Systems Using an Experimental Technique Based on the Jaeger Method, DOI:10.1016/S1359-4311(01)00089-8.

CHAPTER 5

Zhu Ding, Biqin Dong, Feng Xing, Ningxu Han, Zongjin Li, Cementing mechanism of potassium phosphate based magnesium phosphate cement, Ceramics International, Volume 38, Issue 8, December 2012, Pages 6281-6288, ISSN 0272-8842, http://dx.doi.org/10.1016/j.ceramint.2012.04.083.

CHAPTER 6

Maria Chiara Bignozzi, Andrea Saccani, Ceramic waste as aggregate and supplementary cementing material: A combined action to contrast alkali silica reaction (ASR), Cement and Concrete Composites, Volume 34, Issue 10, November 2012, Pages 1141-1148, ISSN 0958-9465, http://dx.doi.org/10.1016/j.cemconcomp.2012.07.001.

CHAPTER 7

Lucia Fernández-Carrasco, D. Torréns-Martín, S. Martínez-Ramírez, Carbonation of ternary building cementing materials, Cement and Concrete Composites, Volume 34, Issue 10, November 2012,

Pages 1180-1186, ISSN 0958-9465, http://dx.doi.org/10.1016/j. cemconcomp.2012.06.016.

CHAPTER 8

Hideaki Yasuhara, Debendra Neupane, Kazuyuki Hayashi, Mitsu Okamura, Experiments and predictions of physical properties of sand cemented by enzymatically-induced carbonate precipitation, Soils and Foundations, Volume 52, Issue 3, June 2012, Pages 539-549, ISSN 0038-0806, http://dx.doi.org/10.1016/j.sandf.2012.05.011.

CHAPTER 9

Chang K-C, Chang C-C, Huang Y-C, Chen M-H, Lin F-H, et al. (2014) Effect of Tricalcium Aluminate on the Physicochemical Properties, Bioactivity, and Biocompatibility of Partially Stabilized Cements. PLoS ONE 9(9): e106754. doi:10.1371/journal.pone.0106754.

CHAPTER 10

Yu L, Li Y, Zhao K, Tang Y, Cheng Z, et al. (2013) A Novel Injectable Calcium Phosphate Cement-Bioactive Glass Composite for Bone Regeneration. PLoS ONE 8(4): e62570. doi:10.1371/journal. pone.0062570.

Index